HOME PLUMBER

AURA
EDITIONS

CONTENTS

Editor: Helen Davies
Art editor: Graham Beehag

Published by Aura Editions
2 Derby Road, Greenford, Middlesex

Produced by Marshall Cavendish Books Limited
58 Old Compton Street, London W1V 5PA

© Marshall Cavendish Limited 1986

ISBN 0 86307 479 0

Typeset in Garamond by Quadraset Limited, Avon
Printed and bound in Italy by L.E.G.O. S.p.a.

While every care has been taken to ensure that the information in *Home Plumber* is accurate,
individual circumstances may vary greatly. So proceed with caution,
especially where electrical, plumbing or structural work is involved.

CHAPTER 1

KITCHENS

If your kitchen plumbing fittings are starting to look a little tired and dowdy, now is the time to replace them, starting with the sink and taps. At the same time, you can introduce a few luxury items such as a waste disposal unit or an electric water heater.

MODERN STYLE TAPS

Many old taps have seen better days, and if they are in kitchens or bathrooms which have been upgraded, they may well now look old fashioned and out of place. Yet the business of replacing them with completely new taps can be daunting simply because it may involve quite complicated plumbing in a confined space.

Tap conversion kits eliminate the plumbing problems. You simply unscrew the head of your old tap and replace its handle and working parts with new ones. It's a simple job which can transform your kitchen sink, hand basin or bath. In the bathroom it means you can co-ordinate the style of all your taps. And because a new mechanical action—the gland—is included with most kits, you will have effectively renewed the life of your tap. The only tool you should need is a spanner.

Practical considerations

You will find that most old taps can be converted—although not all: kit manufacturers only make provision for the most common types. Obviously, the screw threads of the converter and sizes of the parts must be compatible, and it is not immediately obvious how this can be checked. In the UK, manufacturers sometimes specify the British Standards Institute code number (normally BS 1010) of taps for which their kits are suitable. The difficulty is finding the BS code number on your tap. It is usually not visible—although a 'Made in Britain' stamp or British Standards Institute kitemark can usually be found in some unobtrusive position.

The rule of thumb is that if your tap is British made and the kit you are using is British too, then it will almost certainly fit. If you are at all uncertain, write a note to the manufacturers of the kit describing your tap and any marks on it including brand names —they should be able to advise you.

If your taps are suitable, the other main thing to look out for is their size—whether they are bath or basin taps—and make sure your kit matches. Normally bath taps are ¾in while kitchen, laundry and hand basin taps are ½in. In Britain new taps are still mostly reckoned in Imperial sizes and an old tap certainly will be. The new size refers to the diameter of the supply pipe and its connector to the bottom of the tap. So to make certain look under the back of the structure.

Fitting the new gland and head

The first thing to do is to turn off the supply of water to the tap. This may simply involve turning off a stop valve somewhere in the pipe leading to the tap—or it may be easier to turn off the whole system at the mains. But most household plumbing systems are more complicated than that and you will probably have to do some investigation— usually under the sink—with a torch.

If you are converting a hot water tap look for a stop valve between your tap and the hot water cylinder. A cold water tap should have a stop valve somewhere between it and the storage tank in the roof. If you can't find any stop valves you will have to turn off at the mains and drain the tank or cylinder supplying the tap. If you are draining a cylinder remember to turn off any heating element—boiler or immersion heater—for the duration of the job. When you are satisfied that no more water can come from the open tap, give it an extra twist anticlockwise to make sure it is fully open.

If you are very unlucky it may be that there is no stop valve at all inside the house. In this case the supply can be turned off at the water authority stop valve which is usually situated under a metal cover flap somewhere between the house and the road and operated with a key. You may have to ask the authority to do this for you; in some areas you can do it yourself with a piece of wood cut to a V-shaped notch at one end.

With the water turned off, unscrew the bell shaped outer casing of the tap body using an adjustable spanner. The cover should have a thin octagonal or hexagonal rim at its bottom for this purpose. This may need considerable effort to shift if it has corroded in place over the years (see When it won't budge).

Slide the freed casing up the spindle of the handle to expose the hexagonal nut around the base of the spindle. You shouldn't need to remove the handle to do this. Adjust the spanner to fit the nut and unscrew it to remove the mechanical part (the gland body) from the tap. Once again, this may not be easy to do. Pull the spindle to lift away the loosened gland body.

Take this opportunity to clean the base and spout of the tap thoroughly. On chromium plate use a proper chrome cleaner and a gentle action.

Screw in the replacement gland until it is firm, making sure you include any seals or washers. Follow the maker's instructions about final tightness. If the body is plastic— as many are—be careful not to deform it with the spanner.

The new head may be a press fit on the gland body or it may be held with a screw concealed under a pressfit cover. Rotate it until the internal splines on the head line up with the grooves in the top of the spindle. Check that any symbols on the head are the right way round when the tap is closed and —elementary but easy to forget—that the Hot and Cold heads are on the appropriate sides of the tap.

Turn the water on again and check that the taps work properly. Leave them to run for a while to allow any air bubbles to work their way out of the pipes.

When it won't budge

Dirt, corrosion and scale can all make it hard to undo plumbing threads such as those on a tap's outer shroud or gland body.

Before you start to apply any force, take care to prevent the tap from turning. If it is

1 *When turning off water, look for a stop valve—usually under the sink*

2 *You may need to remove the handle and shroud to get a spanner in*

3 *With the shroud off, unscrew the gland unit. Do not use excessive force*

4 *Remove the gland, leaving just a bare casing. Clean the base thoroughly*

5 *Fit the modern style nylon gland in its place. Screw tight with a spanner*

6 *Press on the head making sure that the splines line up with the spindle*

at all loose and turns in its socket, there is a strong risk of damaging the bath or basin. Stop it turning by locking a piece of sturdy batten across the spout.

Use a spanner which fits the nut properly, or a well fitting adjustable one. If it is too short to apply enough force you can extend it by locking another spanner or wrench onto the end.

If a steady pressure doesn't release the nut easily, don't immediately apply more force. There are several things you can try to release the threads first.

Start by trying to *tighten* the nut slightly. This often breaks the grip of the threads so that you can then unscrew them in the conventional way.

If this fails, you can try heat, which will expand the parts slightly. Wrap them tightly in cloth and pour boiling water carefully over the cloth. Try the nut both while it is hot and when cold again.

Penetrating oil is well worth trying, but

Hot water is a great tap loosener

do allow *plenty* of time for it to soak in.

If all these methods fail you can try jarring the end of the spanner to shock the

Use wooden props to stop a tap moving

nut free, but take extreme care on ceramic basins where there is a definite risk of cracking if you use excessive force.

FITTING AN INSET SINK

Old style kitchen sinks, bracketed to the wall or set on top of their own special units, are items that traditionally you don't move. In a small kitchen that has been progressively modernized this often means the positions of the surrounding appliances and units are a poor compromise between the desire for space and the need for efficiency.

The way to solve the problem is by taking advantage of the latest trend in sinks—recessed units that you set into an existing or custom-made worktop.

Inset sinks have one major advantage—flexibility. You don't need a special unit, so you can do away with the existing one and install a straight run of worktop. This could go over extra units to match those you've already got, giving you the added bonus of increased storage space.

Alternatively, you might decide to change the position of the sink to make your kitchen more convenient. And if your worktops are the usual laminated sort, fitting one or more inset sinks in them generally calls for no more than jig-sawing a hole and a few minor alterations to the base unit and waste system below.

Choosing a sink

Most inset sinks are designed to be accommodated within the standard worktop depth of 600mm–610mm. Sink depths generally range from 170mm–200mm. But when it comes to form and accessories, the choice is vast. The most popular combinations are:
• **Single sink**—impractical unless space is tight and you can hang a draining rack above the sink.
• **Double sink**—a better bet, particularly as some models incorporate lift-off bowl covers doubling as chopping boards or drainers. Choose a two-hole mixer tap which you can site centrally between the two bowls.
• **Sink and drainer**—choose a double sink and drainer if you've got the space.
• **Sink and waste disposer**—the waste disposer bowl can be teamed with a single or double sink, but you should allow for the extra depth taken up by the disposer unit itself (this will be sold separately).
• **Sink materials:** The traditional sink materials are stainless steel, vitreous enamel (over cast iron or pressed steel), and ceramic (usually vitrified clay). The last two have been improved enormously in recent years

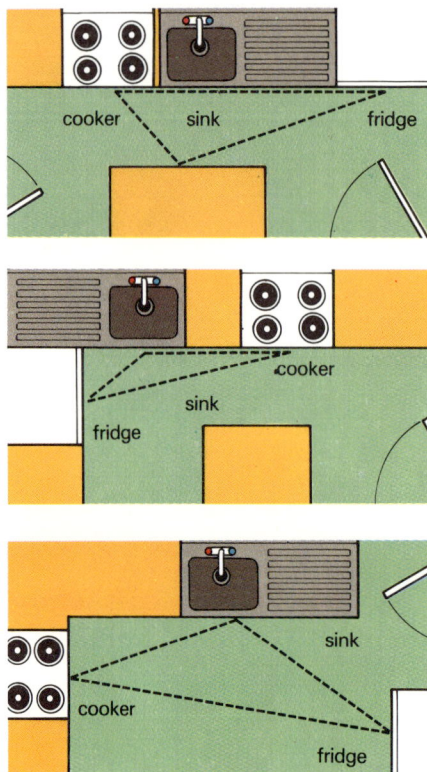

and now come in a wide range of colours and styles to suit most kitchens.

The newcomers are reinforced plastic (ICI's Sylac, Resan) and cast synthetic stone (Du Pont's Corian). These are well worth considering if you want an up-to-the-minute look, although Resan is not 100 per cent heat resistant and Corian, though tough, is a very expensive choice.

Making a choice: This is largely a matter of balancing personal taste, washing up habits and the space you have available against what you can afford. But bear in mind that very shallow sinks are alright for rinsing—not for washing; and that round and oval bowls swap capacity for stylish looks.

Choosing taps: Some inset sinks come with taps (or at least tap fittings); on others it is left to you to fit your own taps direct to the worktop. If you opt for a mixer tap, make sure that the model you choose has water authority approval. Water by-laws usually stipulate that there must be no

Position your sink wisely—it is an integral part of your kitchen 'work triangle'

Sinks come in all shapes, sizes and colours. Some have detachable chopping boards, convenient if you have a waste disposer

waste disposer you might have to sacrifice a shelf as well.

New laminated post-formed worktops are normally 600mm wide and are sold by the metre, though some dealers will supply them cut to size. The worktop can be teamed with ready made non-drawer base units, or units to match those already installed in the kitchen.

If you plan to tile the worktop, do this after you have cut the recess and tap holes but before you fit the sink.

Plumbing and drainage

Once you have decided on the ideal location for the new sink, work out how to get the existing plumbing to the site with as few modifications as possible.

Drainage fittings: What you need here depends on your new sink and on whether or not you're moving it. If you take the latter option your aim must be to stick as close to the original waste pipe run as possible, with no bends. Some local authorities do permit bends in waste pipes providing a clearing eye is fitted on the crown in case of blockages (see diagram). But others don't, and you should avoid them if you can.

If the existing waste pipework is in copper or lead, you may as well replace it completely right back to the stack or gully. The same may apply if your pipes are plastic and you can't match the brand or they are of an odd size—it's not worth risking leaking joints for the sake of buying a few feet of plastic waste pipe.

It's worth getting a waste system manufacturer's catalogue and then sketching out what new parts you need. You may have the choice of push-fit, solvent welded and screw coupled joints, but only the connection at the trap need be semi-permanent. Standard pipe sizes for sink wastes are 38mm and 43mm. Don't forget to bear the following points in mind:

• If you're connecting to an older, open gully with new pipe, buy a plastic grid and carry the pipe outlet below it. If the gully's the modern back inlet sort, you can sever and connect it to the existing pipe above ground level.

• Use the manufacturer's recommended adhesive and cleaner for making solvent welded joints in the pipe run.

Your supplier should be able to help you

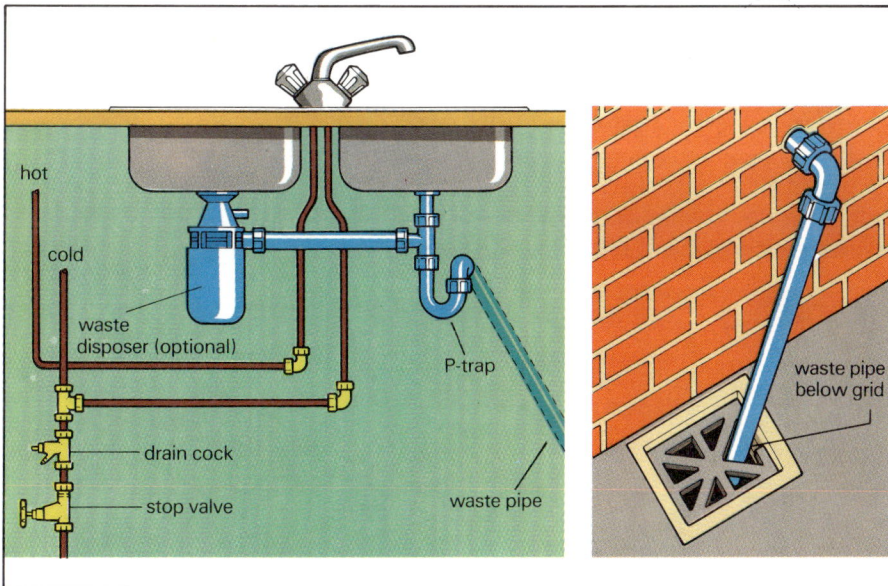

danger of stored hot water mixing with the mains cold water inside the tap—possibly contaminating the supply.

Worktops and base units: If you're fitting the sink into an existing wooden or laminated worktop over drawered base

A typical plumbing layout with the waste run directly to a convenient gully

units, you can discard the drawers, saw off the fronts, and screw or glue and pin these back on from behind. Only in the case of a

with traps and sink fittings—many sinks are sold with their own 'plumbing kit' which includes everything you need in this area. Double sinks normally connect to a single double-sink trap and onto a single waste pipe. Otherwise you have the choice of P or S traps, both of which can be swivelled round to adjust the direction of the pipe run. Avoid bottle traps: although easy to clear, they are also easy to block—especially when used in the kitchen.

Waste outlets are nearly always available from your supplier, although you may have to buy them separately. Most now incorporate a built-in overflow; separate overflows, if you need one, connect between the waste outlet and trap.

Water pipework: If you sever the existing supply pipes near to the old sink and you're not changing its position, it should take no more than a single bendable connector on each pipe to link up to the new taps. Buy connectors with a capillary or compression joint at one end (see below) and a screw tap connection at the other, to match the taps. Remember that British taps still take an Imperial (½in.) fitting, while imported European ones are usually the 15mm size. One hole mixers have two 10mm tails which need adaptors (usually supplied with the taps) to take them up to 15mm.

Even if you are moving the sink, the flexibility offered by bendable connectors—which are 540mm long but can be cut to suit at 180mm intervals—means that you shouldn't need more than one extension length per pipe. Make this up with 15mm pipe and compression or capillary fittings. **WARNING:** The cold supply to your sink will come direct from the rising main, and in the UK some authorities insist that all such pipes are joined with soldered capillary fittings—so before you buy the materials check with your local building control office if you are in any doubt.

Tools and other materials

There are no special requirements here, other than a good supply of plumber's putty, PTFE tape and silicone sealant. Make sure that you have a pair of wrenches, grips or adjustable spanners for pipe jointing—or a blowlamp, if you're making capillary rather than compression joints.

If you don't have a jig saw for cutting the sink recess, it's worth hiring one—the alternative is to use a padsaw, which is very hard work and likely to be much less neat.

Removing the old sink is the easy part of the job, but it still needs to be done with care

if you are to avoid problems later.

Start with the waste pipe as this will give you more room to work underneath the sink. If the trap is plastic, place a bowl underneath to catch the water and then simply unscrew each joint to dismantle the run. If the trap and pipe are metal, cut through the pipe at a convenient place—you'll be replacing them anyway—and dismantle the rest of the run when the sink is out of the way.

> ### ★ WATCH POINT ★
>
> To ensure a square cut, make a mark on the pipe and wrap some thin card around the pipe at this point. Align the edges of the card, then use it as a cutting guide.

Removing an old sink

Now work out where best to sever the supply pipes, bearing in mind that you need enough access to work with a junior hacksaw. If the new sink is going in more or less the same position, cut as near to the old taps as possible; if you are extending the pipes to a new location, cut them at a place where a join will be easy, well clear of the sink and/or sink unit.

Note that if your main stop valve is under the sink, as many are, you must cut **on the tap side** or you'll flood your kitchen.

Shut off the water supply to the pipes next. The cold supply is isolated by closing the main stop valve; the hot by closing the valve on the cold feed pipe at the base of your hot water cylinder or water heater. Empty the hot water pipes by turning on all

the hot taps in the house. Work from the top downwards to avoid airlocks. Make sure that any heating apparatus is switched off. When you cut the supply pipes, make sure you cut them square.

If your old sink is the heavy earthenware sort, support it on bricks before unscrewing or sawing through the brackets holding it to the wall. Get several helpers to assist with removal—old sinks are very heavy.

Newer stainless steel sinks are held to their base units by brackets so you should have no difficulty unscrewing these.

Before lifting the sink away, break any seal along the back with an old screwdriver. If you have a tiled splashback, pull the sink out and up towards you to avoid cracking the tiles.

Installing the new sink

Whether you're modifying an existing worktop, or adapting a new post-formed one, it must be cut to accept the new sink and drilled for separate tap holes.

2 *Use a piece of tape or card wrapped around the pipes as a cutting guide*

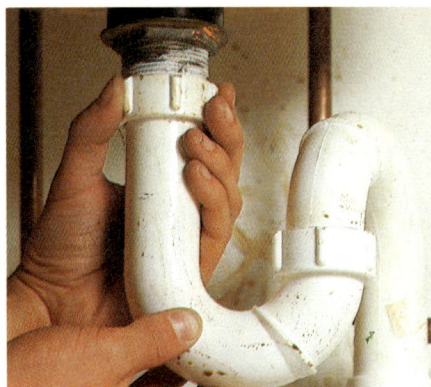

1 *Place a bowl under the trap—to catch any drips—before you unscrew the waste connection*

3 *Remove the seal around the old sink by using an old screwdriver or an old knife*

4 *If no template is supplied, mark the position of the sink on the worktop*

5 *Lay the seal against your mark and draw a cutting line on the inside*

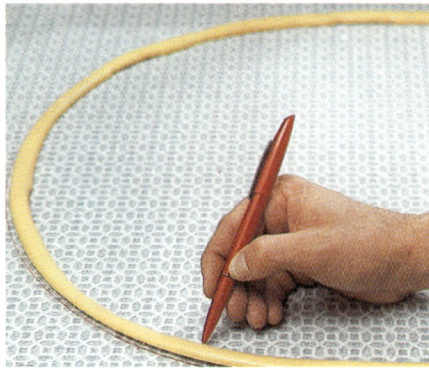

6 *Use a flat bit to drill a hole—large enough for a jig saw blade—on the inside of your cutting line*

7 *Score the cutting line and insert the jig saw blade carefully into the hole. Cut around the marked line*

If the sink is to go into an existing work-top, remove the worktop at this stage by un-screwing the brackets holding it to its base units and removing the screws securing it to adjoining runs.

If the worktop is new, cut and prepare it for installation before embarking on any of the sink modifications.

Manufacturers often provide a template for cutting the sink recess. Use this to mark cutting lines on the worktop, which should be firmly supported on trestles or stools. If no template is provided, mark out the surface carefully with the dimensions given in the instructions. Alternatively, you can mark out the exact position by using the sink itself as a template as shown in steps 4 and 5. Use a try square to ensure that your marking out is at all times square with the worktop edge.

The basic cutting technique for thick laminated worktops is to drill holes within the perimeter of the marked out recess large enough to insert a power jig saw blade. You then saw along lines drawn between each of the holes in turn and finish off any corners with a rasp.

Drilling holes in laminate isn't easy. Either drill pilot holes with a twist drill and then enlarge them with a flat bit, or else drill a series of small holes and join them up. In either case, use a piece of masking tape over the hole marks to stop the drill slipping. And take great care that your holes don't extend outside the marked line.

Before using the jig saw, score along the cutting lines with a laminate cutter.

As you saw, keep the sole plate hard against the cutting guide and let the blade find its own way through the wood—don't be tempted to force it. If it overheats, stop sawing and let it cool down. After you have completed the first line, remove the guide and set it up for the next one—and so on.

With the hole completed, try the sink for fit and make any minor adjustments with a rasp or planer file.

Cutting the tap holes

Position the tap hole or holes parallel with the back of the sink recess. If you're fitting a single mixer, you should locate it centrally behind a single bowl or midway between a double sink.

Arranging the plumbing

Do as much of the plumbing work—inclu-ding equipping the new sink with its fittings—before you install the sink and refit the worktop. Once they're in position, you won't have much room in which to work.

If you're installing new units, start by setting them in place so that you can mark off where the pipe runs must be led through them. Then drill or notch the backs as appropriate.

Preparing the sink: Unless plastic or rubber washers are provided for the pur-pose, you bed the waste and overflow outlets on plumber's putty and then screw them in place by tightening the backnuts from behind. Continue tightening until the putty squeezes out around the rims, then wipe or scrape away the excess.

If the overflow is a separate unit, connect it to the waste outlet at this stage, making sure that you get any washers or seals that come with it in the right order (see diagram). Follow by wrapping one or two

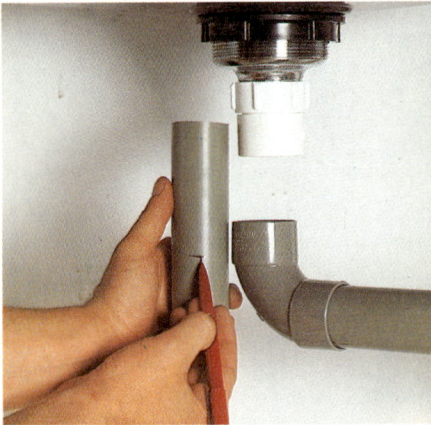

8 *Assemble the waste system in a dry run and measure before cutting*

9 *File off burrs at the end of cut pipes before connecting them up*

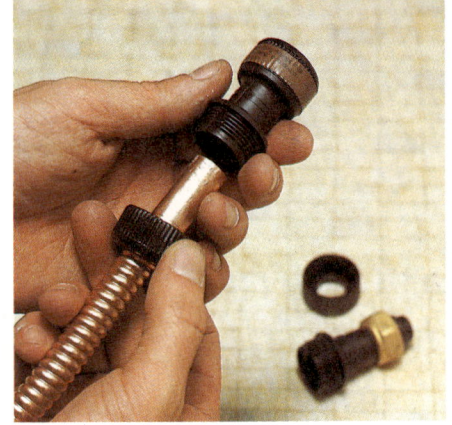

10 *Push-fit joints are the simplest way of connecting the supply pipes*

11 *Washing up liquid acts as a good lubricant on push-fit waste pipes*

12 *On solvent weld pipes, spread the adhesive and join quickly*

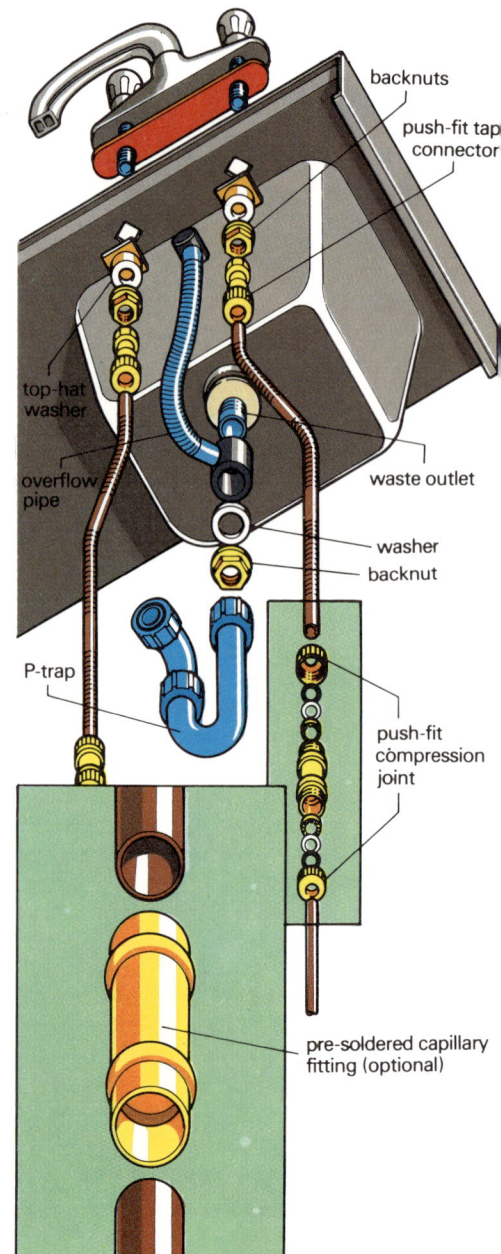

turns of PTFE tape around the waste outlet thread, then screw on the plastic trap—there's no need to tighten it for the moment.

Fitting the taps: If these are separate, they can be fixed to the prepared worktop. Like the sink fittings, they may well be provided with sealing washers; if not, bed them on a generous layer of mastic and trim away the excess when you've tightened up the backnuts. You'll find it easier to use a box spanner or basin spanner for this.

The next stage is to locate the worktop and sink temporarily in position so that you can measure up underneath for the waste and supply pipes. Where the latter are concerned, accuracy isn't essential—you simply make sure that you bring the bendable connectors you'll be fitting to the pipe ends 'within range' of the tap shanks.

It is important, however, to get the waste pipe lined up perfectly with the trap on the sink. If you are replacing the entire run, you'll have some room for manoeuvre; otherwise you need to direct-measure the first length against the trap at this stage.

Remove the sink and worktop. Tackle the supply pipes first. If you need to extend the runs, measure up two extension pieces in 15mm copper pipe, not forgetting to allow for the amount that gets 'lost' inside the joint fitting. Cut the pipes with a junior hacksaw, taking care to get the ends square, then file off any burrs inside and out.

If you're using compression joints, fit them to the severed supply pipes first. Tighten one and a half turns above hand tight and then repeat the procedure on the extension pieces. Finish off by compression-jointing on the bendable tap connectors in the same way.

If you're making soldered capillary joints, see page 11.

Now deal with the waste pipe. If you're joining straight to the existing pipe, measure up and cut a length to run to the new trap. If you're solvent welding the

General arrangement for push-fit supply and waste connection. You may need a straight connection capillary joint to connect the rising main (inset)

backnuts

push-fit tap connector

top-hat washer

overflow pipe

waste outlet

washer

backnut

P-trap

push-fit compression joint

pre-soldered capillary fitting (optional)

joint, try the pipe for fit in a dry run first. Trim as necessary, then file off the burrs, clean the pipe ends, apply the solvent cement and join. On a push fit joint, apply lubricant to the pipe ends.

If you're replacing an entire run, use solvent weld joints. Dry-assemble the run and check the fall—not more than 1 in 48, not less than 1 in 24. When all is well, cement the joints and fill the gap in the wall with non-setting mastic. Complete the run outside to below the level of the gully grid or join it up to a waste stack.

Finally, secure your pipe runs to the walls with brackets—every 500mm for copper, every 900mm for plastic.

Capillary joints

Some authorities insist that you use capillary joints on plumbing jobs involving the rising main. In this case, opt for pre-soldered (Yorkshire) fittings which don't require extra solder.

To make the joints, you'll need a blowlamp, flux, wire wool and a heat-proof board or large tile with which to protect surrounding fixtures and fittings.

As with compression joints, it's vital that the pipe ends being joined are cut square and burr-free. Inspect the fitting beforehand to check that the solder rings inside are continuous—reject it if they're not.

When you are ready, follow the procedure below exactly.
• Clean the pipe end and the inside of the fitting with wire wool until the copper shines. Remove dust.
• Coat the pipe end and the inside of the fitting with a thin but even layer of flux, applied with an old toothbrush.
• Bring pipe and fitting together.
• Adjust your blowlamp flame until the centre is clear blue. Play it gently over the fitting and the pipe end, moving it all the time, until the flux starts to bubble and spit.
• At this point, ease off slightly but keep the flame to the joint; after a few seconds you should see the solder that was inside the fitting appear around the pipe end. When the ring is complete remove the flame and then allow it to cool.

Preparing the worktop

Fit the worktop first, screwing into adjoining surfaces and securing the brackets to the base units as necessary. Follow by laying the rubber seal supplied with the sink in position; in some cases this must be laid on a

Clean the ends of the pipes with wire wool

bed of silicone sealant to prevent drips.

Fit the sink strictly in accordance with the instructions, tightening the clips or brackets that secure it to the worktop.

Working from underneath, connect the bendable connectors to the tap shanks and

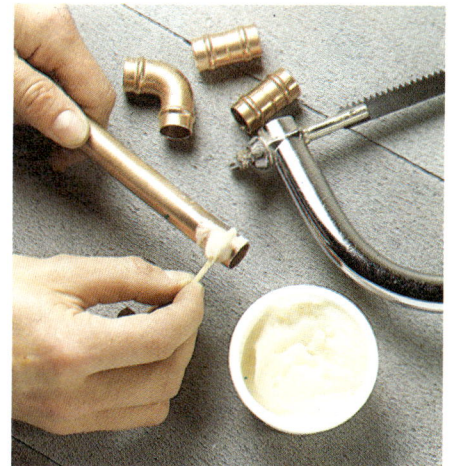

Use a matchstick to spread the flux

tighten with a wrench or adjustable spanner. Finish by push-fitting or screwing on the waste pipe to the new trap and hand-tightening the joints on the trap itself. Reinstate the water supply and test all joints for watertightness before making good.

13 *Secure the worktop to the units with the slotted angle brackets*

14 *Push the seal firmly in place on the underside of the sink*

15 *Tighten the brackets so they grip the under side of the worktop*

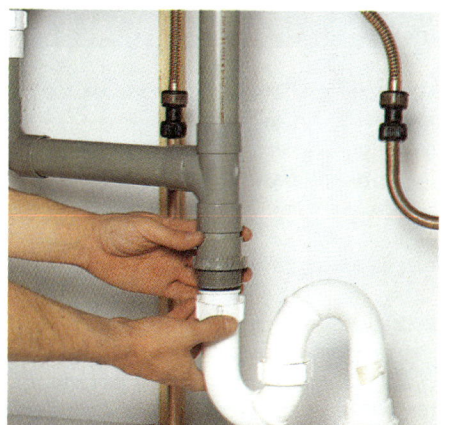

16 *Finally, connect up the waste pipes before turning on the water*

INSTALLING A WASTE DISPOSER

Waste disposal units are the modern answer to smelly rubbish bins; they keep your cooking—and your kitchen—clean and pleasant. Initially they had a poor reputation for reliability but waste disposers have come a long way since then, and today's machines can be relied upon to give many years of safe, trouble-free service—providing, of course, that they're used properly and serviced regularly.

All waste disposers work on more or less the same principle: the waste is flushed through the sink drain hole into a grinding compartment where blades driven by a powerful electric motor convert it into semi-

fluid effluent to be carried away along the waste pipes into the main drain. On most units the grinder is activated by a switch fitted near the sink; on a few more sophisticated models, it comes into operation automatically when the grinding compartment receives a full load and water.

Modern disposers will cope with most kinds of organic waste—fish bones and skin, vegetable peelings and such like—which are unpleasant to dispose of by conventional means. About the only things they can't handle are large bones, tough fibrous stringy items and large amounts of fat, but these are easily thrown away.

The chief enemies of the waste disposer are things made from plastic, metal or ceramic. Nowadays, however, there are plenty of drainage stoppers and rubber trap accessories to prevent items such as cutlery and small bowls from falling into the machine and damaging the blades.

The system

The typical modern waste disposer is fitted under the kitchen sink, suspended from the drain outlet bush by a series of clamps and plates. The machine itself is in two parts: the upper half is the grinding compartment and has a standard 38mm outlet pipe in the side; the lower half houses the blades and motor and has a flex to connect it to the power source. The halves are normally held together by adjustable spring clips or screw bolts, enabling the lower half to be removed easily for clearing or servicing. A blanking plate can be fitted to the top half under such conditions, so that you can still use the sink in the conventional way.

The power can be taken from a conventional 13 amp plug and socket fitted at least 1.5m from the sink, but this is not advisable since it tempts you to operate or even unplug the unit with wet hands. It's far better to install an independent switch fused connection unit as a spur from another socket in the kitchen.

Drainage arrangements are made by connecting a standard U-trap to the unit's outlet pipe, followed by a run of ordinary 38mm piping as far as the stack or gully to which the existing sink connects. If your sink is currently fitted with a bottle trap you need to replace it with a U-trap since it might restrict the flow of what will be slightly denser waste matter. Likewise, if your existing drain run has an old metal trap and lead or iron pipe, this is a good time to replace it with the more modern PVC waste fittings.

Usually the only problems related to drainage occur because the unit's outlet is lower than that of the original sink trap—which in turn affects the fall of the waste pipe. The ideal fall for a waste disposer waste pipe is 1:12 and it may be that you can't meet with this requirement unless

wall socket and plug

stopper

suspension plate

top housing

fused connection unit

grinding blades

waste trap

motor compartment

taps

overflow

swept connector

waste trap

waste disposal unit

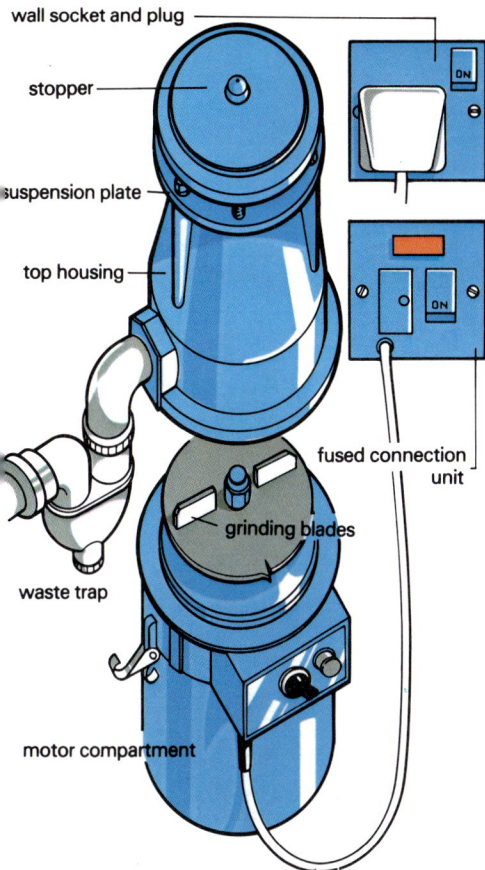

The vital connections: a waste disposer has two halves—the top half is bolted to the sink outlet hole via a series of plates and seals and connected to the waste trap and pipes. The lower motor compartment—complete with grinding blades—is secured to the top with adjustable clips and is connected to the power supply by a flex. The best source of power is via a fused connection unit taken as a spur from the nearest power socket

you lower the entire pipe run by dismantling it and drilling a new hole in the wall. This doesn't normally involve changes to where the waste pipe discharges. But if the outlet is over an open gully, you would be well advised to fit an extension to the pipe and carry it below the level of the grating—if your existing grating is iron, simply buy a new one in plastic and cut a hole in it to allow the waste pipe to discharge safely into the gully.

Sinks and sink outlets

Waste disposal units can be fitted to most sinks—but you should think carefully before making a final decision.

Waste disposers are most useful if they

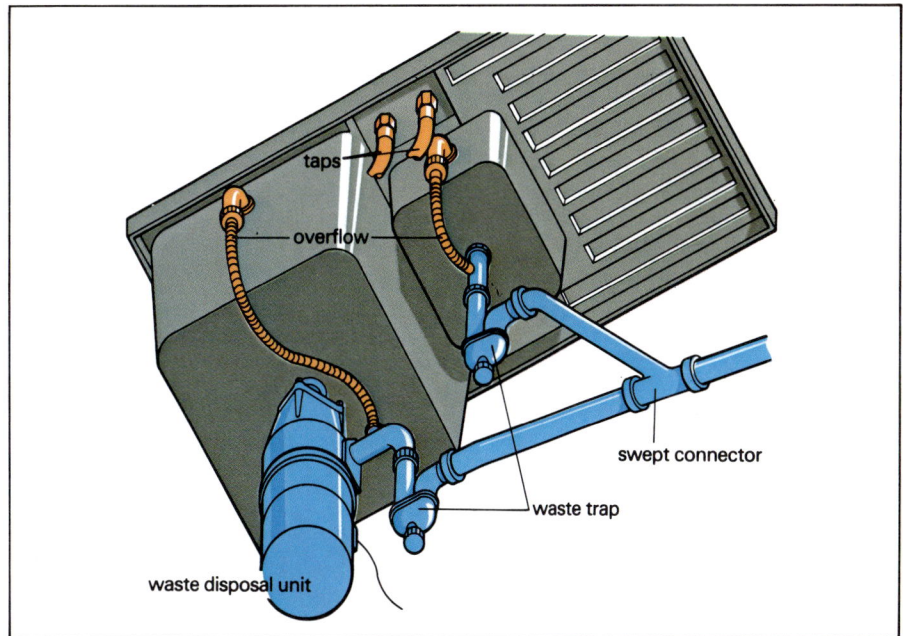

have their own separate sink bowl or, failing that, if they are fitted to one half of a double sink. This gives you a disposal facility while you are actually doing the washing-up.

The standard inlet size for a waste disposer is 90mm, as opposed to a conventional sink outlet's 38mm; the larger size doesn't affect the disposer's efficiency, but it does make it easier to get waste into it, particularly if you're dealing with large amounts at a time.

Most modern kitchen sinks can be ordered with a 90mm outlet in one of the bowls, and many have the format of two large bowls plus a separate small disposer bowl (with a large outlet) to one side (often covered when not in use by a chopping board). Such arrangements are easily plumbed in to a single drain waste pipe using double traps. So all in all, if you're thinking of replacing your kitchen sink it's a good job to do in tandem with fitting a waste disposer.

If you want to fit a waste disposer to an existing sink with a 38mm outlet, you have two alternatives. The first is to take advantage of the fact that most machines are available with special small-outlet adaptors that are no more difficult to fit than the usual 90mm sort. The second is to enlarge your sink outlet to 90mm. This job involves removing the existing bush and buying a new one, then cutting a hole for it using a large hole saw. You also need to hire a special recessing tool to make a recess so that the new bush sits flush with the sink's surface to prevent leaks.

In view of what's involved in relation to what you gain in convenience, there really

Drainage arrangements for a double sink must be planned in advance—you'll need a wider range of waste fittings.

It's best to incorporate separate traps for each basin (right) though you could use the same trap for both outlets.

Waste pipes from the disposer unit must be provided with a fall of at least 1:12 and any changes of direction must be gentle to avoid blockages—swept bends and connectors should be used in preference to right angled elbows.

Keep the arrangement simple to aid servicing and avoid blockages

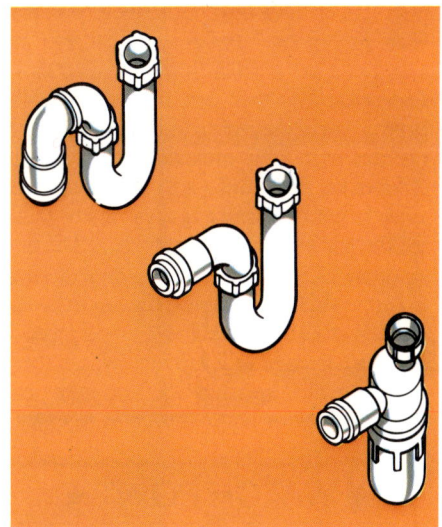

Waste traps come in a range of shapes. Bottle traps (bottom right) must be replaced with S-traps (top) or P-traps (centre) otherwise the denser effluent may cause blockages

is little point in taking the second option—it's far better to fit a new sink with a disposer compartment or at least with a 90mm outlet.

Nearly all types of sink are strong enough to take a waste disposer and to absorb the slight vibrations made by the unit when it's operating. The only exceptions are some thin plastic and aluminium sinks or old ceramic sinks—consult the sink manufacturers if you are in any doubt—and it is possible to add an extra support made from offcuts of board.

Planning ahead

This is one job where it pays to buy the actual unit some time before you plan to install it. Doing so gives you the chance to check what modifications, if any, must be made to the existing drainage layout and you can also take advantage of any specific recommendations which are made by the manufacturer of the unit.

With waste disposers, you get what you pay for. Most models now feature rubber jaws on the inlet boss to guard against accidental entry of unwanted objects, plus some sort of stopper should you want to use the sink in the normal way. The better units on the market have an automatic overload cut-out to stop the motor overheating and a reverse-run facility for clearing minor blockages. You will, of course, pay more for automatic operation.

When you buy a waste disposer it should come complete with outlet pipe, flex and its own special sink boss. But before you part with any money, make sure that you are also given the correct fitting for your sink.

Electrics: For this part of the job you'll need a surface-mounted switched fuse connection unit with indicator light, enough 2.5mm sq twin and earth PVC sheathed cable to connect it to a nearby ring or radial socket (see The power supply, page 16), and a supply of plastic conduit (mini-trunking) to carry both the cable and the flex from the unit. Surface mounting is generally much easier to arrange in a kitchen (where there may be a tiled splashback). You may need a flex connector and extra cable to extend

the length of the flex—bear in mind that the colour code is different for cable than for flex.

Drainage: Start your assessment of requirements here by inspecting the existing layout. Metal pipework and bottle traps must be replaced with a U-trap as a matter of course. If you already have a plastic U-trap, place a bowl underneath, unscrew it from the sink and waste pipe, then test-fit it on the disposer outlet pipe: if it is a non-standard size, it too must be replaced, together with the pipe run.

The next thing to check is the fall of the waste pipe. Offer up the complete disposer unit into position and check how much lower the outlet is than the existing waste pipe. You should be able to tell by eye if the pipe needs lowering. In this case, if the joints are solvent welded, you'll need to sever the pipe and buy new sections to deal with the pipe where you make a new hole in the wall; if the pipe joints are push-fit, re-use the old rather than buying new ones.

After your inspection, make a sketch of the proposed new pipe run and note down the extra parts you need. If you're buying a new trap, make sure it has the same fittings—overflow connection, washing machine inlet, and so on—as your existing trap. If you're connecting the disposer into part of a new double sink arrangement, you'll need parts to make up a trap layout like the one shown in the diagram on page 13.

Apart from the electrical and drainage materials, you need a supply of non-toxic plumber's putty to fit the new sink boss. Installing the disposer itself should call for no more than a wrench and a screwdriver. You will, of course, need more tools—drill, masonry bit, mains tester screwdriver, hacksaw, wire strippers—for the wiring.

Preparing the sink

This part of the job is straightforward enough—it's mainly a matter of removing existing fittings—but if you're going to re-use the existing waste pipe and trap, do not disturb them more than necessary.

Start by placing a basin under the existing trap. If it's a bottle or U-trap, you can simply unscrew it by hand; metal traps must be severed on the waste pipe side with a hacksaw and then disconnected from the sink boss by unscrewing the retaining nut with a large wrench. Replace metal pipework with new PVC pipe and fittings.

Now turn your attention to the sink boss itself. Some are loosened by unscrewing the retaining nut from below, using a wrench,

1 *Remove the existing waste trap. Drain first to avoid a mess. Bottle traps must be replaced with more efficient S, P or U-traps*

2 *Release the sink boss retaining nut using a large wrench. Considering the options, now is a good time to change your sink for the type with a wider 90mm outlet hole.*

then prising or gently tapping upwards to break the putty seal. Others are in two halves, which you separate by unscrewing the screw in the centre of the outlet from above—again prise or tap to break the seal. With the boss removed, scrape or scour away all traces of old putty from around the recess and from below the sink.

If you need to lower the waste pipe this is a good opportunity to make a new hole in the side of your sink base unit. The easiest way of doing this is with a hole saw, set to about 40mm. Alternatively, mark the hole using a spare piece of waste pipe as a template, drill a series of holes around the circle, then join them with a padsaw.

Installing the unit

Though nearly all waste disposers are connected to the sink in a similar way, the actual connection method on your unit may vary slightly in detail from those given below—always consult the manufacturer's instructions for your particular model.

Start by separating the two halves of the unit. Take the upper part, note carefully how the various plates, clamps and seals are arranged, then dismantle them.

Prepare the sink outlet by laying a generous layer of non-toxic putty around the boss recess, then slip the new boss into the correct position.

Most 90mm connections follow the sequence shown in the diagram opposite.

3 *Before you attach the new sink boss, spread a layer of non-toxic plumber's putty around the boss recess*

Using standard 38mm outlets (right) means fitting an adaptor plate to the sink bush. The adaptor plate is held in position by extra sink bush retaining nuts. Apart from the inconvenience of the narrower outlet, the waste disposal unit is just as efficient in use

Wider sink outlets—90mm—are specially designed for waste disposers (below). The unit attaches to the bush by way of a series of plates and seals—the top is bolted to a suspension plate. Grub screws force the clamp plate against the seal

Working from underneath the sink, slip the sealing gasket over the boss, followed by the clamping plate. Next, offer up the suspension plate, locate the grub screws on it with the corresponding holes in the clamp plate, then hold it in position by snapping the steel circlip supplied onto the boss thread (on some models there is a screw-on retaining nut instead of a circlip).

The next stage is to tighten the grub screws on the suspension plate so that the clamp plate and the gasket above it are forced hard against the underside of the sink to create a watertight seal. This done, offer up the top half of the unit (check that its seal is in position around the lug on the inlet), locate the lug in the boss and the retaining bolts in their holes in the suspension plate, and slip on the nuts and washers provided.

Hand tighten the nuts, then swivel the unit until the outlet pipe points in the desired direction. When you're satisfied with the arrangement, tighten the bolts fully to pull the unit hard against its seal.

If you are using a 38mm outlet, the procedure is slightly different. First of all, secure the sink bush by slipping on its seal

and then tightening the retaining nut from below. Follow by screwing on the suspension plate upper retaining nut — about two

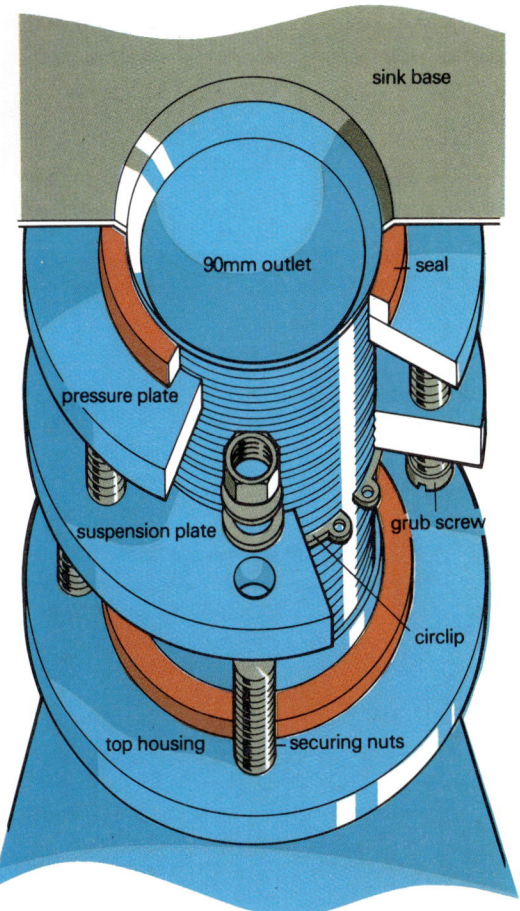

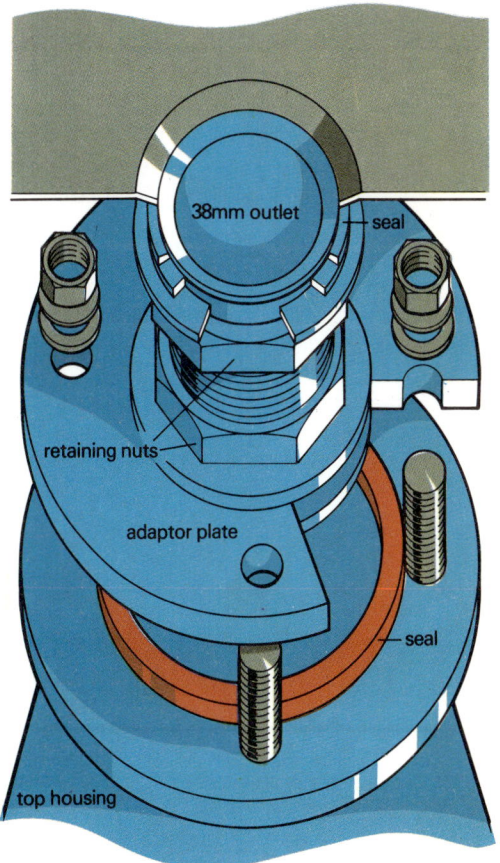

sink base

90mm outlet

seal

pressure plate

suspension plate

grub screw

circlip

top housing

securing nuts

38mm outlet

seal

retaining nuts

adaptor plate

seal

top housing

plastic grid and cut it to accommodate the pipe. If all is well make any permanent joints that are necessary by pulling the joints apart, deburring and degreasing the pipe ends; then apply solvent cement and reconnect.

If you haven't had to alter the pipe run, simply mark off against the trap how much pipe has to be removed, trim it with a hacksaw, then juggle it into the trap joint to make the appropriate connection.

In all cases, resecure the pipe run to the wall with clip brackets every half metre. Seal around any new holes with mastic, inside and out. Double check that all the joints are secure, especially those either side of the trap and on the unit itself.

4 *Fit the sink bush and slip the seal, clamp plate and suspension plate into place. A steel circlip completes the arrangement of this particular design. Tighten the grub screws substantially in order to form a tight seal*

5 *Fit the waste outlet pipe to the top half of the unit, then slip the bolts of the housing into the holes of the suspension plate. Adjust the position of the waste disposal unit so that it suits your drainage layout*

The power supply

These instructions are for installing a fused connection unit—by far the safest way of getting power to your new waste disposer.

thirds of the way up the remaining thread on the bush.

Now slip on another seal, the suspension plate itself, a further seal, and finally the lower retaining nut. Tighten this nut with a wrench to hold the plate firmly. Then simply bolt the upper half of the unit to the plate to complete the job.

Arranging the drainage

What's involved here depends very much on the location. The critical part of the job is ensuring that the waste pipe ends up with the recommended 1:12 fall.

If you think you can use the existing pipe as it stands, start by loosening all the wall brackets holding it in place.

Screw or push-fit the U-trap to the disposer outlet pipe, then hold the waste pipe against the trap and check the fall. If it is out by any appreciable degree, you'll have to lower the entire run of the waste pipe.

In this case, dismantle the existing pipework as far as the drainage outlet. If the pipe joints were glued, cut through them squarely using a hacksaw and a piece of card as a template. Now make up a new set of pipework in a dry run to extend from the trap as far as the wall and adjust the fall using the method described above. When it is correct, mark where the pipe must pass through the kitchen unit or the outside wall (or both).

Make the hole in the wall by working from both sides in turn and using the old

6 *Fit the new trap to the outlet pipe from the waste disposer. Use PVC fittings and run the pipework to join up to the existing waste fittings—in most instances, the new pipes will be much lower than the old*

hole as a reference point. It is good practice to use a piece of larger-diameter pipe as a channel for the new pipe: cut it to the thickness of the wall and set it in the hole with exterior filler. Block up the old hole at the same time using repair mortar or filler.

Now reassemble the entire pipe run in another dry run, cutting and joining it where necessary. Lead it through the wall and join it to the trap at the sink, then recheck the fall. If you don't already have the facility, arrange for the waste pipe to discharge into the gully *below* the grid—buy a

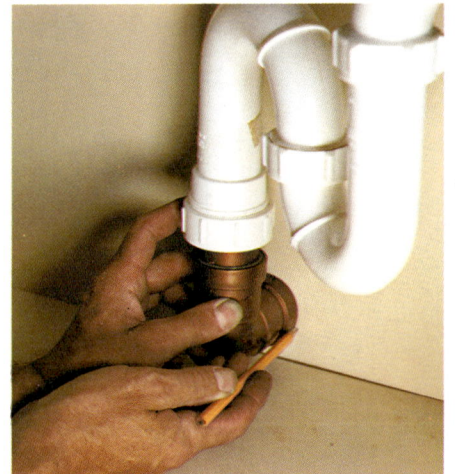

7 *Attach new pipe sections—with bends or sweeps—to the waste trap. Try to lead the pipes as near as possible to the old waste fitting—mark the base units so that you can cut a new exit hole*

Fused connection units are almost like extra sockets—and they're installed in the same way, by taking power from another socket. If you have not already done so, select a site for the unit: it can go above or below the sink, but if you put it above, it should be out of arm's reach of anyone standing directly at the sink, and at least 150mm above the height of the kitchen worktop surface.

Make a start by unscrewing the body of the unit and offer up the backing plate to the wall—check that there's enough flex from the disposer itself. Mark the screw pos-

8 *Drill a series of holes around the marked exit route for your waste pipe. Cut out the waste with a padsaw or trimming knife and saw blade*

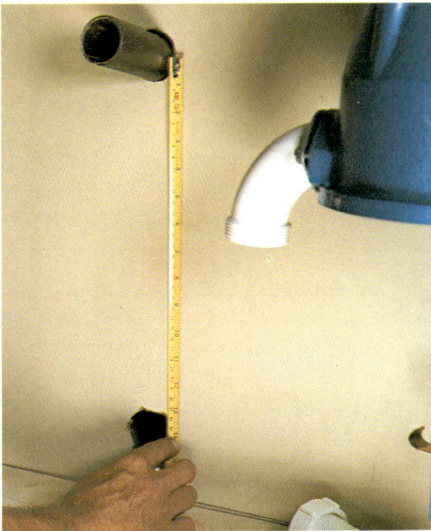

9 *Measure the distance from the old exit hole to the new. Use this dimension to cut a hole through the wall for the waste pipe. Use a drill and masonry bit or a cold chisel*

itions, then fix it in place. If the flex is too short, fit a flex outlet plate.

Now turn off the electricity at the mains switch on the consumer unit or fuseboard.

Unscrew the faceplate of the socket you propose to take the power from. If you have a ring main system, the back of the socket reveals two sets of wires (red, black, and green and yellow PVC sheathed) and you can safely take power from it. But if you find three sets of wires, the socket already has a spur drawn off it and you can't connect into it. If the socket has only one set, it is itself a spur. In this case it's inadvisable to connect

to it without first seeking the advice of a qualified electrician.

If you have a radial system an inspection of the socket will reveal one or two sets of wires. If there is only one, you should be able to connect to it providing there aren't too many sockets further down the circuit —but consult an electrician if in doubt. If there are two sets, it is not advisable to connect to the socket.

Connecting the unit: Once you have found a suitable socket, arrange a run of plastic conduit between here and the new connection unit. Fix the backing with screws and plugs, masonry pins or impact adhesive (whichever is most convenient) and use angle pieces where changes in direction are required. Afterwards prise the existing socket's backing box away from the wall, knock out a new cable entry hole and —unless it's surface-mounted—fit a rubber grommet to protect the cable.

Prepare the ends of your 2.5mm sq cable, lay it in the plastic conduit and snap on the conduit cover pieces. Sheath the bare earth wires with green and yellow sleeving and connect the cable to the appropriate terminals on both the socket and the connection unit faceplate (often labelled MAINS)—red to live (L), black to neutral (N) and bare wire to earth (E)—as shown in the diagram on page 18.

Having arranged an exit point for the cable into the conduit, replace the socket faceplate.

Connecting the waste disposer: Fit the lower half of the unit to the upper half following the manufacturer's instructions. Now feed the flex up to the connection unit.

Loosely pass the flex to the fused connection unit by the most convenient route— along the base units of the kitchen and up through the worktop is generally best. Avoid stretching the flex. If it is long

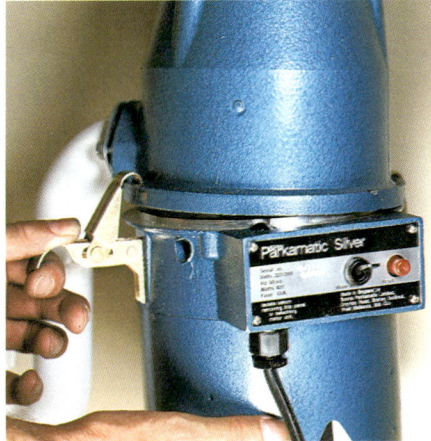

10 *Fit the motor compartment to the top housing of the unit. Ensure that the seal is in place, then adjust the clips to form a tight seal*

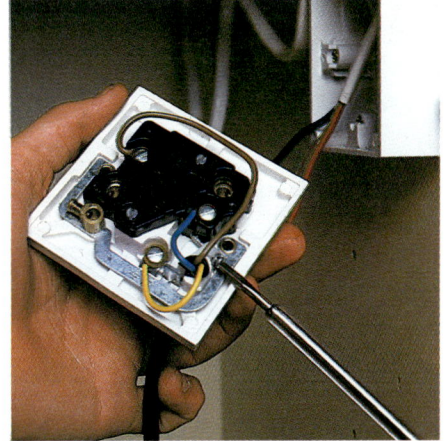

11 *If the flex is too short to reach the fused connection unit, fit a flex outlet plate underneath the sink. Connect the flex to the LOAD—tighten the cord*

12 *Connect new cable to the FEED side terminals, noting that the terminals are different for cable and flex*

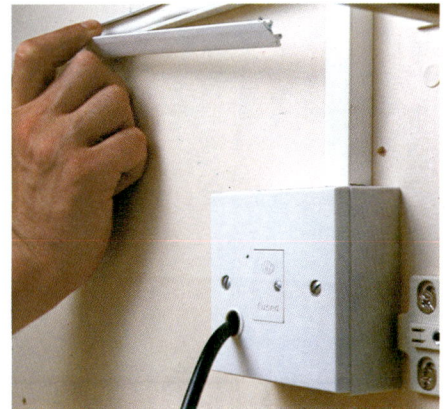

13 *Protect the cable inside surface-mounted plastic conduit—nail, screw or glue the conduit to the walls*

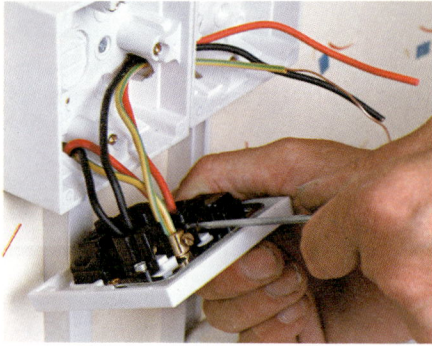

14 *Switch off at the mains. Connect new cable from the flex outlet to the LOAD side of the fused connection unit*

15 *Make a spur connection between the fused connection unit and the nearest suitable power socket (see diagram)*

enough to reach the connection unit, connect it in the manner shown below. Knock out an entry blank in the backing plate and connect to the terminals marked 'LOAD' as follows: brown to live (L), blue to neutral (N) and green/yellow to earth (E). Make sure the flex is held in the flex grip, then secure it with cable clips.

If the flex is not long enough to reach the connection unit you must join the flex to some new 2.55m sq twin and earth cable to extend its length. You need a flex outlet plate for a safe, reliable join between the flex and the cable.

Use an unswitched flex outlet plate with a surface-mounting box and fit it out of the way underneath the sink somewhere—as far from the water supply as length allows. Connect the flex to the 'LOAD' side and the new cable to the 'FEED' side in the manner described for fitting direct to a fused connection unit. You must then join this new cable to the fused connection unit in the manner shown in the diagram below. Cable colour codes are as follows: black for neutral (N), red for live (L), unsheathed copper for earth (E). Sheath the earth with green/yellow PVC sheathing.

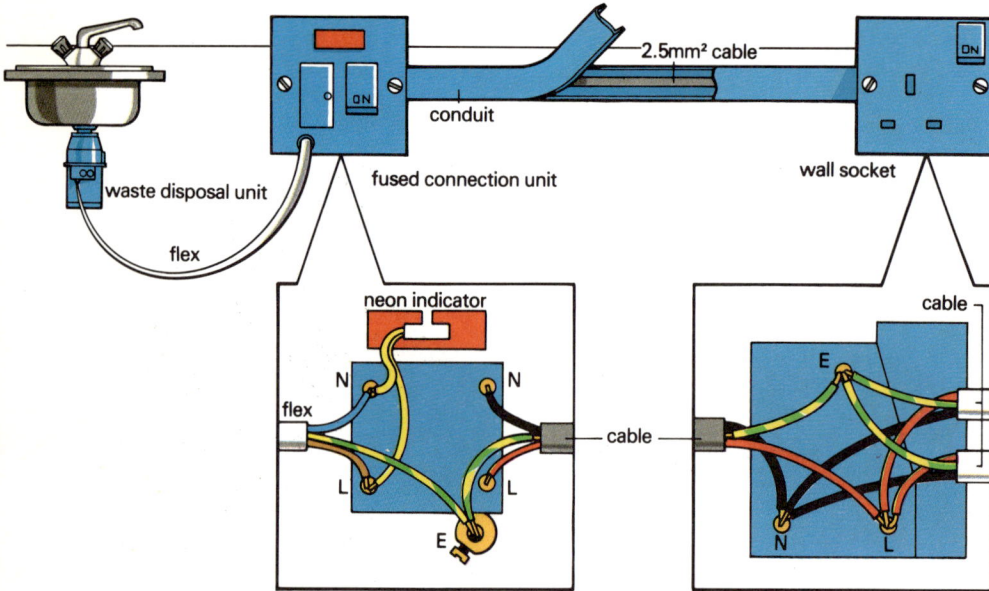

Ideal connection to the power supply: the flex from the unit connects to a fused connection unit at least 1.5m from the

sink. Power comes from the nearest socket—make a spur extension and protect the new cable inside conduit

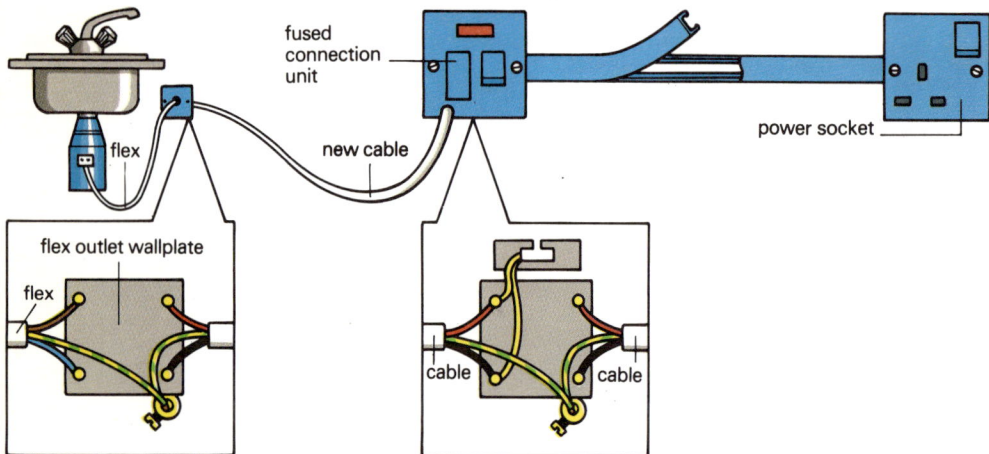

Where the flex will not reach comfortably to the fused connection unit, extend the flex via a flex outlet plate under the sink.

Electrical connections are similar to page 17 except at the fused connection unit—cable joins to cable here

Using a waste disposer

Waste disposers need to be treated with respect if they are to work properly so follow the guidelines below to make sure that you get trouble-free service from your unit.

• Always operate the disposer with a strong flow of COLD water. Never use hot water: it can melt fats in the waste, which then solidify and block the drain pipe.

• Never force waste into the disposer with your fingers or any other object—use a jet of water instead.

• Keep the stopper on when the machine is not in use, to reduce the risk of objects accidentally falling in.

• Always turn off the power supply to the machine before dismantling it or attempting to unblock it by any other means than reverse winding.

• If the cut-out operates, be sure to investigate the cause of the trouble and leave the machine for a while before pressing the reset button.

• When grinding, keep the machine running and the water flowing until you hear only the sound of the motor and swirling water.

• On no account feed the disposer with inorganic materials, large bones, scallop shells or large quantities of fat—these will only damage or block the machine.

CHAPTER 2

BATHROOMS

Even a large scale job like plumbing in a bathroom suite is made a great deal easier if you use plastic wastes and fittings designed with the do-it-yourselfer in mind. And there is no need to stop there. You can easily fit a bidet, shower, shower tray or towel rail using exactly the same techniques.

INSTALLING A NEW SUITE

Bathroom fittings have improved enormously in the last decade, with more styles, colours and choice of fixtures available than ever before. Your existing bath, basin and WC may function perfectly adequately, but swopping them for a new set could give your bathroom that special individual touch it probably lacks. And more importantly, an up-to-the-minute bathroom adds significantly to the value of your home.

If you're far from convinced about your ability to tackle the job yourself, take heart: there's no denying that it's a major undertaking, but thanks to the range of plumbing fittings and attachments now made specially for the do-it-yourselfer it's not half as difficult as it sounds. Such fittings add a little to the cost, but the saving in effort is worth it in the long run.

Planning your approach

Removing the old bathroom equipment may seem the easy part of the job—it is—but before you rush in and start pulling out the fittings, avoid potential headaches by planning your approach.

Firstly, lay aside ample time to complete the job and, if need be, get a friend to help you out: an extra pair of hands can save a lot of time removing the bath and disposing of the rubbish. A day to remove the old suite and two days to install the new one should give you some leeway, and allow for inevitable stoppages which are bound to occur.

Secondly, assess your bathroom with a view to how the new suite, when you buy it, is going to fit in. By siting the new bathroom equipment exactly where the old used to be, there should be no need for extensive modifications of the existing plumbing. Inevitably, you will have to reposition the pipework slightly, as the new suite will not match the old. But modern accessories—flexible copper pipe for tap connections and adapters for waste outlets—make this task simple by comparison with running new drain and supply pipes.

If you have a high wall-mounted WC cistern which is fed from above by a pipe that comes through the ceiling, you will have to remove the pipe and install a new feed. But this too is relatively simple and is covered in detail on page 26.

If you discover that your existing plumbing is made of or includes lead, now is the time to replace it with modern materials—a job best left to a professional. It is not a good idea to join lead pipe to copper because the water flowing through will encourage electrolytic action—like a car battery—and cause one or the other to decay rapidly.

Before you buy your new suite, make a sketch plan of the existing fixtures and a floor plan (on graph paper) of the room as it stands. As well as the overall dimensions, measure the height, length and depth of the bath, basin and WC, and mark them on the drawings. Then consider if you want the new equipment to be larger or smaller, bearing in mind that space is often at a premium in bathrooms.

It is also helpful to make a sketch of how the WC is connected to the soil outlet (see the diagram below) so that you can ensure you get a similar fitting on the new unit. Slight variations in the angle and dimensions of the WC outlet can be overcome by fitting an adapter, but try to get the new WC trap as close as possible to the existing outlet for ease of connection.

Your old WC pan will have one of the outlets illustrated above. Whichever type it has, you should be able to buy plastic adapters to connect your new WC pan to the old soil pipe. Make a sketch and take measurements of your particular outlet.

Most modern WC pans have horizontal outlets (top) which are connected to the soil pipe with adapters. Some old styles have S-trap outlets which go vertically through the floor (centre). There is a third type, P-traps (bottom), which go out through the wall to be discharged into a soil stack on the outside of the house

Choosing a new suite

Bathroom equipment is continually being updated and re-styled, but assuming that you know what you want there are still practical considerations to be borne in mind.

Baths are normally made and sold in one of three materials: acrylic, pressed steel and cast iron.

Before—an efficient but rather dingy bathroom. The high wall-mounted WC and old-fashioned suite give the room a 'dated' atmosphere

After—simply replacing the suite and redecorating gives the bathroom a new lease of life and makes plumbing connections easy

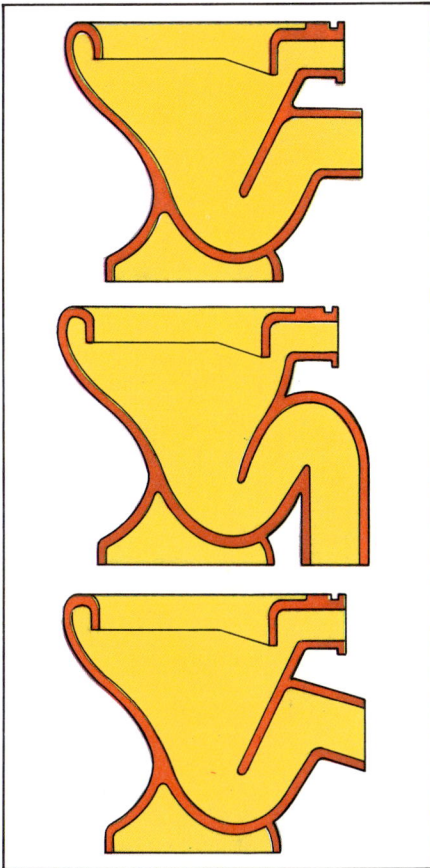

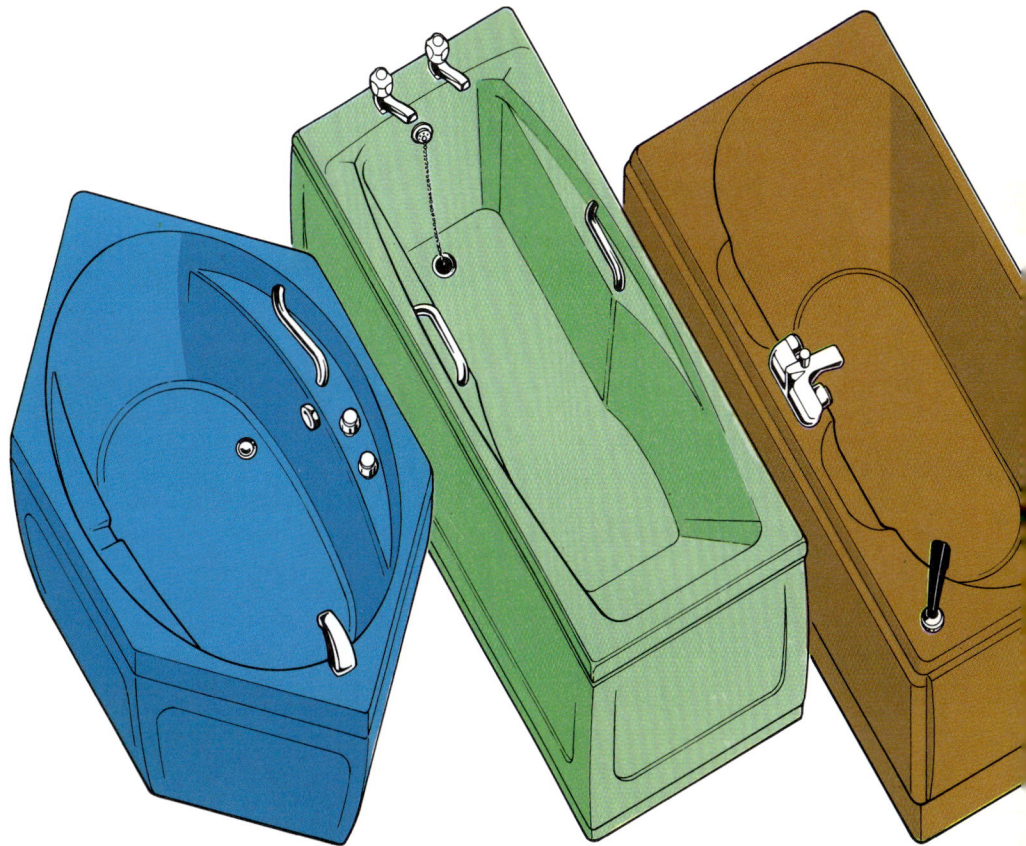

There are plenty of modern bath designs worthy of consideration before you make your final choice.

Corner baths (left) add elegance to a bathroom as well as being more economical on space than the traditional rectangular bath (centre). But, don't forget that rectangular baths can vary both in length and width, and some have sloping backs for greater comfort.

The position of the taps on the bath is also worth thinking about—taps on the side of a bath are more accessible (right) than the conventionally placed ones.

The choice in bathroom suites is very large and so you should take a good look around before making a final decision.

Acrylic baths are supported in a cradle which is assembled around the bath before installation. They are light, easy to fit and are comparatively cheap to buy. Panels are normally included in the kit, although they may be offered only as an optional extra.

The problems once associated with acrylic—creaking and poor heat retention —have now virtually been eliminated; in fact, modern acrylic retains heat better than most other materials. Their chief drawbacks are that they are easily scratched—particularly during installation—and are vulnerable to high temperatures, so take care if you are using a blowlamp near one.

Acrylic baths are available in a variety of shapes—round, square, rectangular and corner-shaped. But a word of warning: before buying a shaped bath, consider the plumbing side of the job first and check that the tap and drain connection points are easily adapted to match your existing pipework and wastes.

Pressed steel baths are also light and easy to install but they are not available in such a range of shapes. The other snag with pressed steel baths is that they are easily chipped and damaged if knocked accidentally with something hard.

Cast iron baths are very heavy and are extremely cumbersome to install, though they do have a feeling of superior quality. These baths stand on adjustable legs and it is important to spread the weight of the bath by placing wooden battens under the feet (it is a good idea to do this on all baths, but essential with cast iron). The disadvantages of cast iron baths are that they conduct heat away from hot water more quickly than others, and they are expensive.

Basins are almost always made from glazed china, which makes them vulnerable during installation. Most modern basins stand on a *pedestal* which takes the weight while the basin itself is held in position against the wall by either brackets or screws. The pedestal also serves to hide the feed and drain pipes.

Wall fitted basins release the floor space

below but they must have a strong wall to support them. On a frame wall, this means removing a section of wall boarding and adding a timber noggin—say 150mm × 50mm—where the fixing will go.

Vanity basins are incorporated into a framework which is boxed in to provide storage space below and room for sundries on either side. Some basins come complete with units, others have compatible units available as optional extras. If you don't like what's available, bear in mind that making your own will take up extra time.

Basins, like baths, are available in a range of styles and shapes but again beware: although some styles look very attractive,

Choose your basin from the wide selection available. The standard pedestal basin (below left) is very stable when in position and is a safe bet. Square and oval basins (below) make a striking alternative to the normal shape but make sure that they are practical to wash in as well as good looking. Basins are usually made from vitreous china so take extra care during installation.

Close coupled WCs (near right) may be relatively more expensive than the traditional low level WC (centre right) but they are quieter and look neater. WC pans can be shaped (far right) for a more streamlined effect.

Although you can get plastic cisterns, most, like the pans, are made from vitreous china—so take care

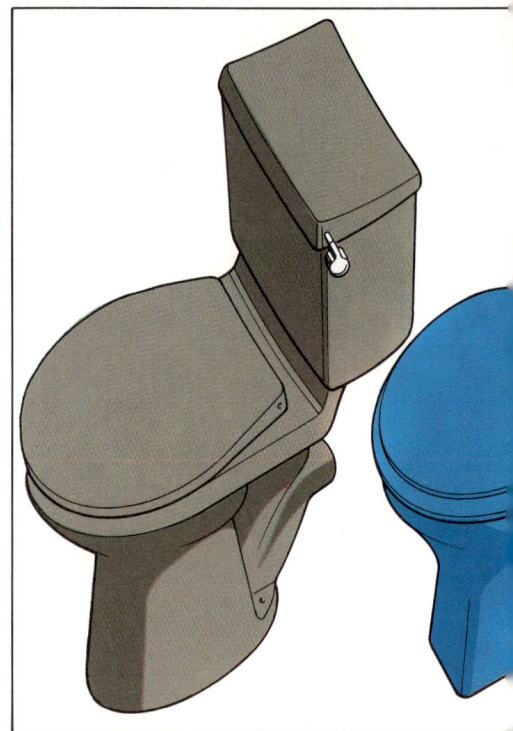

they are not always practical to wash in.

While considering the bath and basin, think about the taps and outlets that are to go with them. These may have to be bought separately, though you should be able to purchase them at the same time. Mixer taps are an obvious option but they cannot always be plumbed into a cold water supply coming directly off the rising main. If you are in any doubt about whether or not you can fix mixer taps, consult your local water authority.

Pop-up waste attachments make the old fashioned plug-and-chain obsolete. There are no problems involved in fitting pop-up wastes—so long as your basin is designed to take them—and full instructions are given when you buy them.

WCs are made up of two parts; the pan and the cistern. Both are usually made from vitreous china although you can now get lightweight plastic cisterns.

With WCs, you get what you pay for. The most basic type has a *low level* cistern which is connected to the pan via a *flush pipe*. Although this does its job perfectly well, a more attractive—and expensive—alternative is the *close coupled* type which is bolted onto the pan. The most quiet, efficient and expensive type of WC employs what is called a *siphonic flush system*.

The water supply to the new WC cistern can be fed from underneath or from either side—a point worth thinking about when you make your choice, remembering that your existing plumbing has to be connected

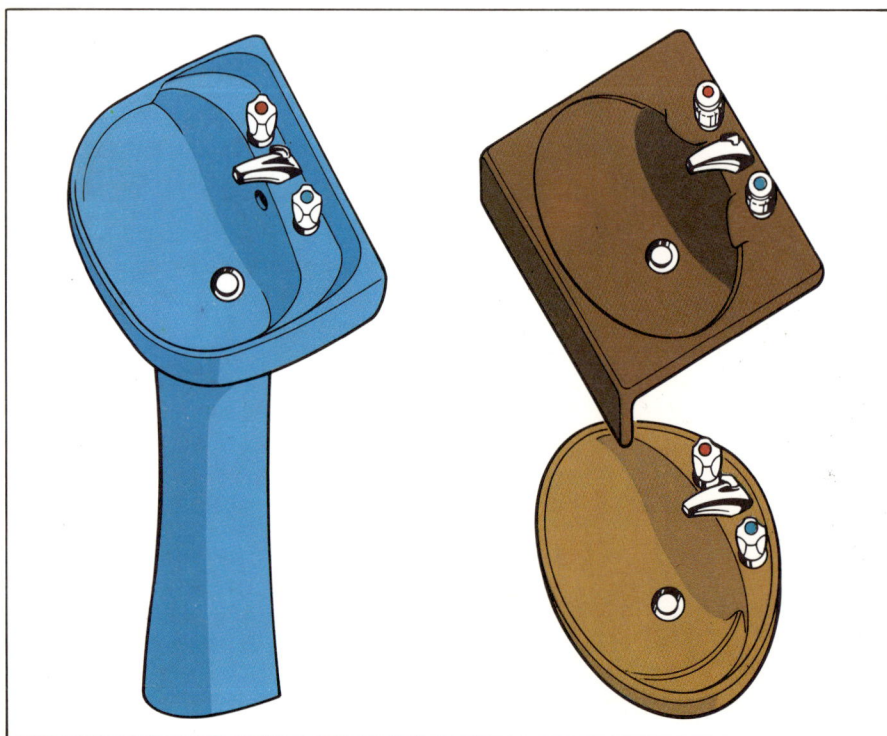

to it in the most convenient way. If your old cistern was fed from above and you intend to install a new feed, it is obviously neater to have a cistern that is fed from below. The overflow and inlet connection points on either side of new cisterns are interchangeable, so you won't have to re-route your plumbing from one side to the other.

Tools and materials

It's impossible to give a comprehensive tool list for a job like this: so much varies according to circumstances and the chances are you'll end up using every tool you've got. However, make sure that you have a hacksaw to cut through pipes, an adjustable spanner for compression joints, a basin spanner to remove taps (see diagram right), a long screwdriver for taking out the securing screws on the WC pan and a tape measure to measure up the old and new equipment and pipes.

You may also find it useful to have the following: a crowbar to lever the old WC pan if it proves obstinate; an old wood chisel to remove any tiles or sealant that secure the bath to the wall; a pipe wrench to unscrew ancient waste outlets and steady pipes; a club hammer and cold chisel to chip away at the WC soil pipe seal; a stepladder or hop-up to reach a high mounted cistern, if you have one; goggles and gloves if you have to break up a cast iron bath; and a sponge to mop up water in the WC cistern.

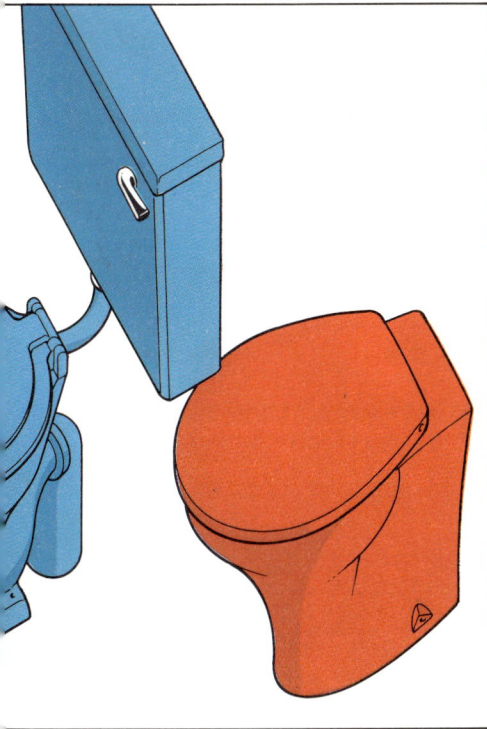

To help remove the taps and other fittings, you must have a supply of penetrating oil; the type which comes in aerosol tins is easier to apply in awkward corners.

If you are unable to complete removal and installation on consecutive days, or if you want to use your water supply elsewhere in the house at the halfway stage, you can temporarily cap the severed supply pipes with a soldered or compression cap. By doing this you can be sure that the rest of your water system will be perfectly safe.

You will certainly have a lot of debris and rubbish to dispose of once you have removed all the old fixtures—so get a supply of plastic rubbish bags and arrange to take the old suite to a local rubbish tip.

★ WATCH POINT ★

Use a basin spanner to unscrew the tap fixings. You can get better leverage by gripping the bottom with an adjustable spanner.

Preparing the bathroom

Even if you intend to keep the bathroom decor as it is, temporarily remove as many wall fittings—mirrors, cabinets and so on—as possible before the upheaval begins.

Which fixture you remove first largely depends on how your bathroom is arranged: you may, for example, have to remove the basin in order to get at the WC. However, once the bath is removed there will be a lot more space so all things being equal this is probably the place to start.

Before you tackle any pipework, turn off the hot and cold water supplies and switch off any water heating apparatus. The supply

pipes may well contain stop valves to isolate them locally but these are not always completely reliable: it is usually quicker and easier in the long run to shut off the supply at the cold water storage cistern (if you have one) or to close the main stop valve and open up all your taps for a few minutes to drain everything down.

The first thing to do is remove the bath panel. This may be made from any number of materials—from hardboard to marble.

Panels are usually secured underneath the lip of the bath at the top and screwed to battening at the bottom. To get at the screws you may have to pull away skirting: this is easily done with a crowbar or claw hammer, levered against a block of wood. If you don't intend to salvage the panelling, breaking it up on the spot will make it a lot easier to carry away. If pipes pass through the panelling, take care not to wrench them out of position as this could dislocate the joints. Where necessary, provide extra clearance by cutting away part of the panel with a padsaw.

Once the panelling has been removed, you have access to the feed and waste pipes. Rather than unscrewing the taps at this stage, it is easier to cut through the feed pipes—a maximum of 300mm from the taps—with a hacksaw: the gap between the sawn off pipe and the new tap connections can be bridged later with copper bendable connectors. If you decide to keep the existing taps, they can be removed more easily with a basin spanner once you have lifted the bath out.

If the waste is connected to a lead pipe, your best bet is to cut through it and renew the piping all the way to the waste stack or gully: lead has a limited life span and it will have to be renewed sooner or later. Plastic waste pipes and traps are easily unscrewed and may be worth keeping to fit to the new bath if they are in good condition.

If the waste pipe is made from iron or copper, and providing it is 38mm in diameter, you can buy a plastic connecting unit which allows new plastic pipe to be jointed onto it.

Remove the waste overflow at the bath or at the trap—whichever is the most convenient to get at.

Some baths are anchored to the wall with brackets and screws, so check this possibility and if necessary remove them before attempting to move the bath itself. If you want to keep any tiling around the bath, unscrew the legs and lower the bath onto some baulks of timber so that it is free of the sealant or grout around the edges. If the tiles are not important, lever them off with an

1 *Remove panels with an old screw-driver levered on a block of wood*

2 *Cut squarely through the tap supply pipes with a hacksaw*

3 *Alternatively, unscrew connections with a spanner or basin spanner*

4 *Disconnect the overflow from the bath by unscrewing the locking nut*

5 *The locking nut on the waste pipe will probably be inaccesible so cut through it with a hacksaw*

6 *Lift off any quadrant tiles or sealant around the bath with an old screwdriver levered on scrap wood*

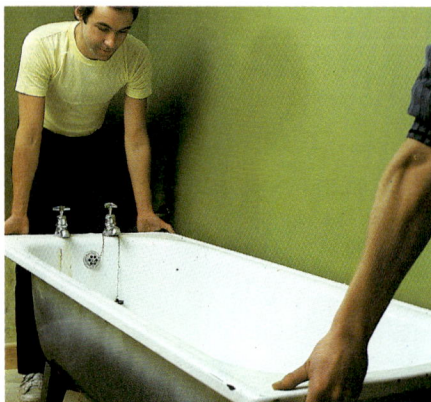

7 *Acrylic or pressed steel baths are quite light but are awkward to handle —so get a friend to help you do the removals. Cast iron baths are very heavy and must be broken up on the spot. But take necessary precautions first: protect your eyes with goggles and lay an old blanket over the bath to contain splinters of enamel*

old wood chisel and then chip away the sealant. If there are quadrant tiles around the bath, they will have to go.

Cast iron baths are best broken up on the spot as they are so awkward and heavy to move. Cover the entire bath with an old blanket to contain splinters, protect your eyes with goggles, and start breaking off small pieces with a sledge hammer, beginning at the edge and working inwards.

Both acrylic and pressed steel baths are light and can easily be carried away by two people—they are awkward to manoeuvre though, so pad the corners with rags.

★ WATCH POINT ★

When you sever the supply pipes, make sure that you cut them squarely to save jointing problems later.

Removing the basin and WC

The procedure for removing a basin is similar to that for a bath. Again, it is much easier to sever the supply and waste pipes near the fixture and then strip it of its fittings (if you need them) on the ground.

If the basin is wall mounted, chip away any sealant around the edges and then locate the fixings, or bracket fixings, on the wall. Undoing them is likely to prove difficult, even if you can get sufficient purchase on the screw heads. You may need to loosen them by driving in a screwdriver or old wood chisel between the basin/brackets and the wall itself. This may well cause the wallplugs to pull out, so make sure the basin is supported from underneath at this stage. Having removed all the fixings, lift the basin clear.

Deal with the fixing on the basin part of a pedestal basin the same way, then chip away any cement around the joint between the two parts. Having lifted the basin clear, you should have more than enough room to tackle any screws holding the pedestal to the floor—you will need a large screwdriver to shift these. Finally, remove the pedestal which can then be disposed of.

Taking out a WC sounds like a really unpleasant task. In fact it isn't that bad, but plumbers will nevertheless charge a fortune to do it for you.

Start by flushing the cistern to empty it, then mop up any water left inside with a sponge. Disconnect the supply pipe at the aid of your assistant before removing the brackets themselves from the wall.

Lastly, pull whatever remains of the old flush pipe out of the pan.

Removing the pan: If the pan itself is joined to the soil pipe via a polypropylene sleeve, or the pan trap is held in the soil outlet collar by mastic/putty, you are lucky: having removed the pan fixings you simply pull it away.

However, it is more than likely that the trap is cemented into the outlet collar, in which case you have to smash the trap—NOT the outlet socket—with a hammer and cold chisel. Wear protective goggles when you do this.

Afterwards, stuff a large piece of rag down the open soil outlet to stop debris falling down and smells rising up; make absolutely certain that there is no chance of the rag itself getting lost down the pipe. Don't bother to chip away the rest of the pan trap debris and seal from the outlet collar at this stage: the job is best left until later, when you know how the new WC pan is going to join it.

With the old fixtures removed and disposed of, you quite literally have to pick up the pieces. Now is the time to make good any defects revealed by their removal.

Holes in plasterwork will be obvious, but you may have more problems with a

8 *Unscrew the basin trap and waste pipe. Watch out for spillages*

9 *Cut through the supply pipes at right angles with a hacksaw*

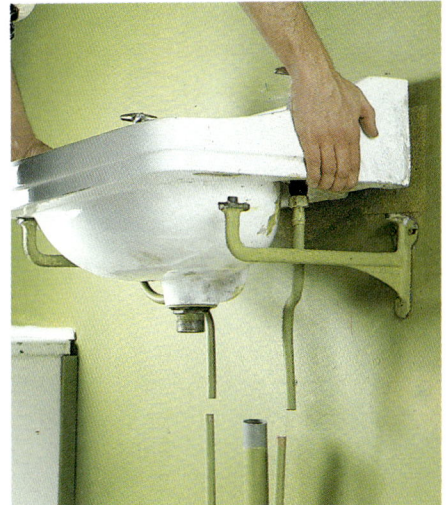

10 *You can then lift the basin free from its brackets or fixings*

ball valve by undoing the connector nut with an adjustable spanner, then do likewise for the overflow pipe.

You disconnect the flush pipe where it joins the siphon unit at the base of the cistern. There should be a large nut holding the pipe to the siphon outlet, and you can undo this with a wrench. But if the flush pipe is an old lead or iron one, you can simply saw through it.

Before starting to remove the cistern, support its weight as best you can. Low level ones should present no problems, but for a high level one you may need steps and a helper standing by to be sure of avoiding any accidents.

Search right around the cistern until you have located all its fixing screws. If it is screwed directly to the wall, remove the screws using the largest screwdriver you've got, then lift it up and away from the site.

If the cistern is on brackets, disconnect them and lower the cistern to the floor with

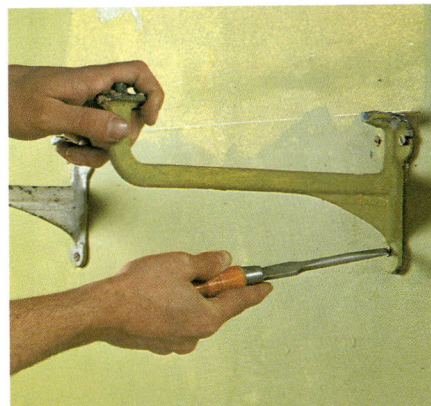

11 *Unscrew the basin supports from the wall taking care not to pull out any plugs or damage the wall*

Now undo the fixing screws holding the pan to the floor. If they prove reluctant— or if the joint has been sealed—use a crowbar to lever the pan free. Finally, remove it from the site with the aid of your helper.

12 *Spray penetrating oil (such as WD40) on the tap fixings and then unscrew them with a basin spanner*

wooden floor. Check all the boards—but especially those where the bath was—for defects and rot. If you suspect dry rot, call in a specialist immediately and be prepared to have to halt proceedings.

Removing the WC

1 *Use an adjustable spanner to unscrew the cistern inlet. A high level wall-mounted cistern is shown here*

2 *Disconnect the flush pipe from underneath the cistern. An adjustable wrench should free it*

3 *Pull the flush pipe away from the WC pan. In most cases this will be a rubber connector*

4 *Use a long screwdriver to remove the securing screws. There will be one screw on each side of the base*

5 *Break the pan trap with a sharp hammer blow if it cannot easily be removed in one piece*

6 *Chip out any residual debris from the soil pipe. Stuff a rag down the open soil outlet*

1

2

3

4

5

6

Capping a pipe

Capping severed copper supply pipes is a useful way of making the rest of your water system serviceable in the interim or permanently sealing a pipe you may have had to cut in the roof space. Apart from compression caps themselves—½in. or 15mm for basins and WC supplies, ¾in. or 22mm for baths—all you need are spanners, flat and round files, and some jointing compound or PTFE tape.

After cutting the pipe, file off the burrs, inside and out, and slightly chamfer the outside edge. Push the securing nut over the end of the pipe followed by the olive—the brass ring—and smear some jointing compound over the end of the pipe. Push the cap on as far as it will go and screw the nut hand-tight. Give the nut an extra one and a half turns with an adjustable spanner to seal the joint completely.

Installing the new fittings

Careful planning is the essence of a job like this: once you start, you don't want to have to stop to make a dash for the plumber's merchant because something doesn't quite line up. Nor do you want the family's washing and toilet facilities out of action for any longer than you can possibly help.

New suite: As soon as this is delivered, check that the fittings which go with it are all there.

• The WC should have a lid—although you may have to buy this separately—and fixings. Inside the cistern in a bag you should find the handle, ball valve assembly, siphon unit, overflow outlet and various nuts and washers. There should also be an instruction leaflet telling you precisely what goes where.

• Acrylic baths are generally supplied with a frame (which you assemble yourself), wooden carcase (to which you attach the panel) and a moulded side panel. Fittings are separate, but you should have ordered them at the same time. Remember that you need an overflow, drain outlet and taps.

• With a basin, again check that you've got the taps, drain outlet and overflow. If your taps incorporate a pop-up waste, check that the fitting instructions are there.

If possible, put the new suite in a room near the bathroom and assemble the fittings in there—it avoids confusion.

New plumbing: This is the most difficult part of the job, so it pays to take as much time over it as you can afford—installation is easy if you are using the right parts.

Although it's laborious, the safest way is to locate each part of the new suite in turn in its proposed position. You can then take accurate measurements or draw sketch plans of how to bridge the gap between new fittings and old pipework.

• With the WC, your main problem is getting the new pan trap to line up with the old soil outlet. The connection will be made using one of the patent sleeve connectors now available for this purpose ('Multikwik' is the best known brand), and these can take up slight variations. Offset connectors will cope with slight differences from side to side. And the 90° type easily convert a modern horizonal outlet into the older P type outlet.

However, if the new pan is too low, you must build it up to the level of the soil outlet with a platform of plywood, blockboard or chipboard. Stand the pan on your chosen material, draw around it, then cut the board to shape and finish the edges with a planer file or sandpaper.

If you have a problem with the WC connectors it always pays to take a sketch plan with you as well as accurate measurements: plumber's merchants usually have a fitting to cope with most common difficulties.

• With the bath and basin, drainage traps

are the main consideration; it's worth replacing your existing ones if they are metal (vital if the new bath is acrylic) and it generally makes the job simple anyway because you can 'tailor' the new trap to match your exact requirements.

You have a choice between conventional U and P traps and bottle traps—the latter are easier to unblock but discharge water more slowly. Often it comes down to what fits best. U and P traps have interchangeable halves, providing for numerous variations.

Traps are available to match all sizes of outlet pipe (note this down before you buy) and come with a screw connector to the waste fitting on one end and a compression or push fitting to the waste pipe on the other. There is normally no restriction on depth, except that in the UK a trap connecting to a combined soil/waste stack must be the 75mm (deep seal) type in case of any back-siphonage.

Any remaining gap can be bridged with a length of compatible plastic waste pipe, compression-jointed into the existing pipe. If you have an old waste system in metal pipe, see Joining to old pipework.

• Before measuring up for the supply pipes, decide what plumbing system to use. Copper is still the most widely available. And if you use flexible tap connectors on all the connections, this could well be all the fittings you need.

However, you may find it pays to cut back the existing supplies and run new ones by a slightly different route. Again you could use copper—with push-fit or compression joints and fittings. Or you may prefer plastic: some systems are flexible, enabling you to clear obstructions easily; others rely on fittings for changes in direction. You have the choice of push-fit or compression joints both of which have their pros and cons.

One thing is certain. If you plan the route carefully enough and take accurate measurements, you should be able to rule out the need for traditional plumbing—blowlamps, pipe bending and so on—altogether.

Bear in mind that baths (unless fed directly from the mains) take 22mm pipe; basins and the WC take 15mm pipe. One hole basin mixer taps have 10mm flexible connectors which can be joined to the supply pipes by means of compression jointed adapters.

Before you order the supply pipes, consider the material you are connecting to—usually either copper or lead pipework. If the existing pipes are Imperial copper, a 15mm compression fitting can go straight

on to ½in. pipe but a 22mm fitting needs a special olive to connect a ¾in. pipe. If your existing pipes are galvanized iron you cannot join copper to them—use plastic instead, with the relevant adapters. If your supply pipes are lead, it's probably worth replacing the entire run—at least to the main feed from the cold storage tank.

Coping with disruption: If you only have one WC, consider hiring a chemical one so that you aren't rushed into completing. These are available at hire stores and sometimes from your local authority health department. You may also be able to avoid disruption by capping off pipes temporarily (see page opposite) or by planning to install the WC first, straight after the bath.

If you need to do the installation work in stages, start with the bath: this is bulkier than the other fittings, so putting it in first lessens the chances of damaging it and gets one large component out of the way.

Specialized equipment

Your requirements here are certain to vary according to circumstances. You will of course need a supply of wrenches and adjustable spanners for the fittings and supply pipes, but you may also need to hire one or two specialized items.

If your existing WC soil outlet is vitrified clay or cast iron you will probably have to cut it to fit. This is best done with an angle grinder or cutting wheel attachment for your power drill. And if you need to make a new hole in the wall for the WC overflow, hire a heavy duty masonry bit—and possibly an industrial drill—at the same time.

Make sure that you have supplies of both PTFE tape and plumber's putty (such as Plumbers Mait)—they are used practically everywhere.

Check too that you have enough fixing

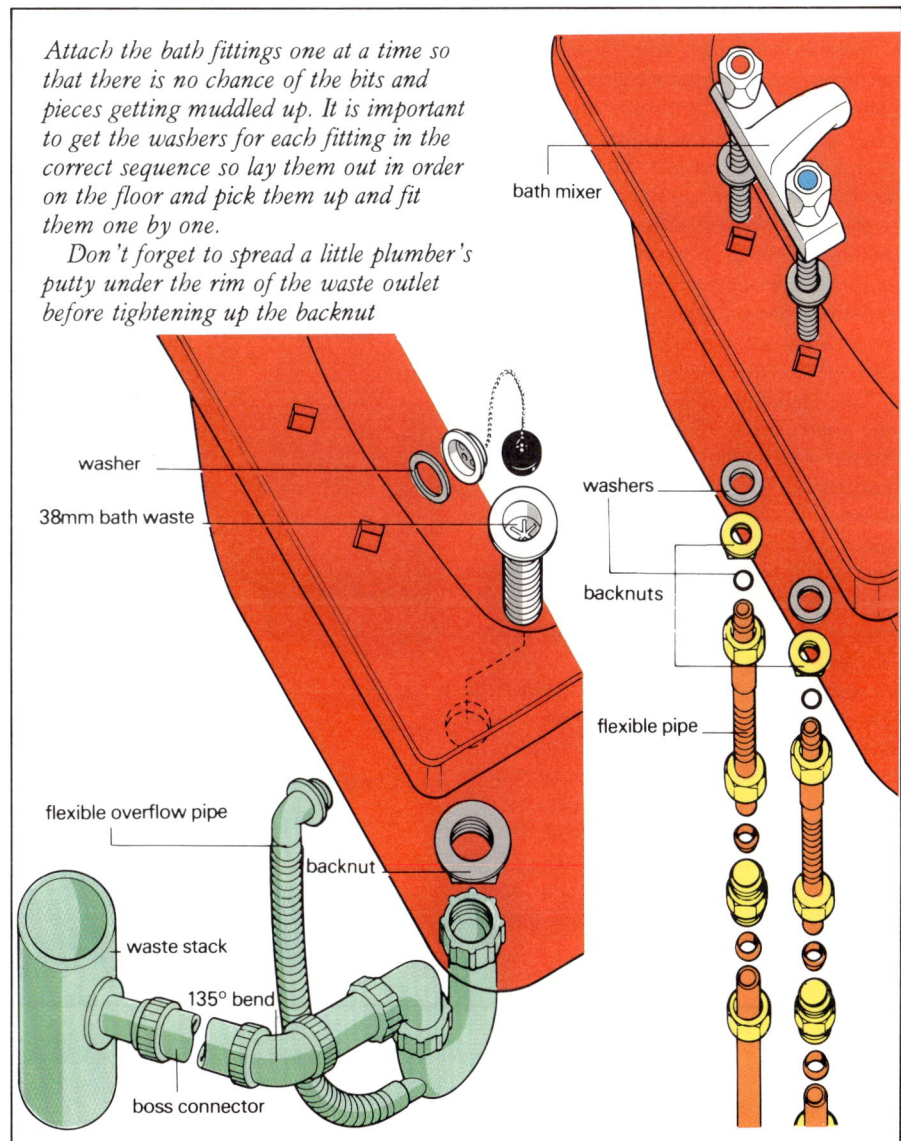

Attach the bath fittings one at a time so that there is no chance of the bits and pieces getting muddled up. It is important to get the washers for each fitting in the correct sequence so lay them out in order on the floor and pick them up and fit them one by one.

Don't forget to spread a little plumber's putty under the rim of the waste outlet before tightening up the backnut

washer
38mm bath waste
flexible overflow pipe
backnut
waste stack
135° bend
boss connector
bath mixer
washers
backnuts
flexible pipe

screws and wallplugs to fit the new suite. Recommended sizes should be given in the manufacturer's instruction leaflets.

Preparing the suite

Equipping the bath, basin and WC with their relevant fittings is best done before installation. And if you add the tap connectors as well, you may be able to cut down on a lot of fiddly plumbing later.

Start with the bath taps. Whether single or mixer, the fitting procedure is much the same. Bed the taps or tap unit on the rubber seals provided—if none are, lay a bed of plumber's putty—insert the tails through the holes in the bath, and slip on the sealing washers from underneath. Follow with the backnuts, which you can tighten with a basin spanner or box spanner.

Now screw a 22mm bendable connector to each tap tail and tighten with a wrench. Leave the connectors hanging straight down.

The overflow will probably connect to the waste outlet via a flexible plastic pipe. Spread a little plumber's putty around the underneath of the waste outlet rim and insert it in the bath. From underneath the bath, slide on the overflow pipe collar (making sure it aligns with the hole in the outlet), then the O ring seal, and then the plastic locknut that secures the fitting. Tighten gently with a wrench.

Spread more putty around the overflow plate, insert it in its hole, and screw on the free end of the overflow pipe. To tighten the plate, you can hold it with two screwdrivers arranged in a 'crossed swords' pattern.
Pop-up wastes: The fitting procedure here will vary slightly from the one above, but full instructions are always given. Make sure that everything is properly bedded.

Installing the bath and basin

If your new bath is acrylic or pressed steel, assemble the frame around it according to

1 *Use a basin spanner to tighten the bath tap backnuts*

the manufacturer's instructions. Lift the whole assembly on-site with the aid of an assistant if possible.

The bath will have an in-built fall, so the sides must be level. Check this with a spirit level along both the length and the width and adjust the frame feet acordingly. Then crawl underneath and mark the wall bracket fixing points on the wall. Remove the bath, drill and plug the holes, then replace the bath and secure the fixings.

2 *Bed the overflow washer in putty before tightening up*

Basin fittings are secured in much the same way as those for a bath. Bottle traps take up less room than the conventional type and are ideal for basins.

The sequence of washers is important, so make sure they're in the right order. Screw up the backnuts hand tight and then give them another 1½ turns with a spanner—don't be tempted to overtighten them as they could well damage or crack the basin

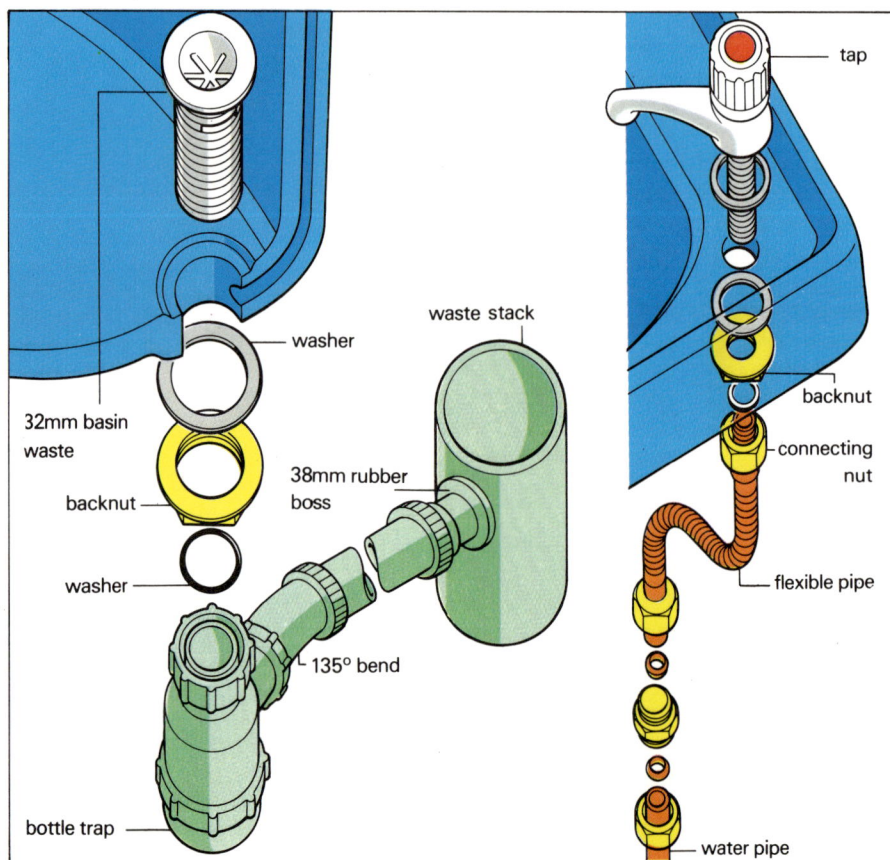

waste stack

washer

32mm basin waste

backnut

washer

38mm rubber boss

135° bend

bottle trap

tap

backnut

connecting nut

flexible pipe

water pipe

3 *Attach the wall brackets securely to the frame battens*

4 *It's as well to fit flexible connectors to the bath and basin taps at this stage*

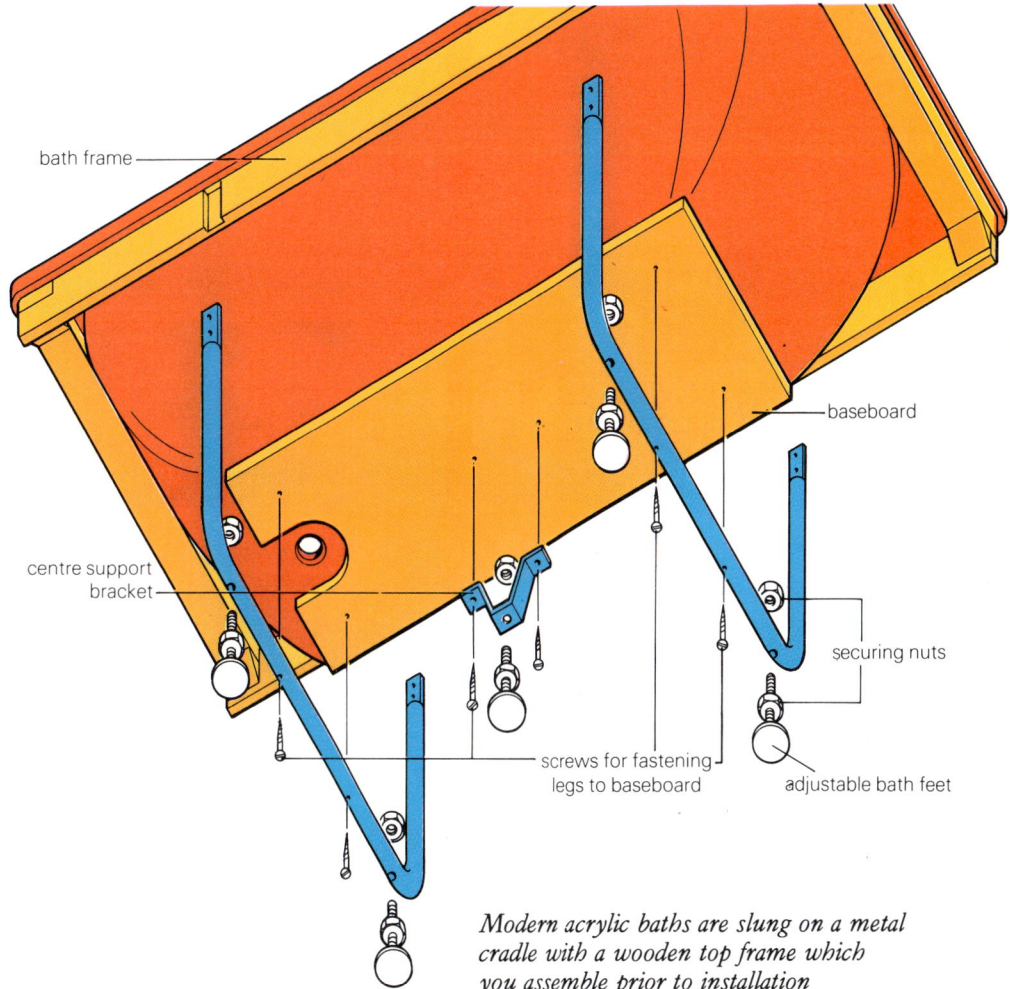

bath frame

centre support bracket

baseboard

securing nuts

screws for fastening legs to baseboard

adjustable bath feet

Modern acrylic baths are slung on a metal cradle with a wooden top frame which you assemble prior to installation

If your bath is cast iron, prepare four offcuts of board to place under the feet. Locate these on-site, then get one or more assistants to help lift the bath into position. Cast iron baths are very heavy, so you're certain to need help. Level up as above.

Fit the trap next, wrapping PTFE tape around the waste outlet thread before screwing on the connector. You should now be able to assess what needs to be done to bridge the gap to the existing supply and waste pipes.

Unless it's going to obstruct you unduly, you should install the basin next, then deal with all the supply pipes at once.

On a pedestal basin, start by positioning the pedestal and mark its fixing positions on the floor. Drill pilot holes, then screw it securely in place.

Locate the basin on top of the pedestal and mark the wall, then refit the basin and secure (not forgetting the putty in the socket trick—see Watch Point).

For a wall mounted basin, mark the bracket positions on the wall at the desired

height—get an assistant to help you, so that you can check that they are level and plumb with a spirit level. Drill and plug the wall and secure the brackets. Then place the basin on top and secure it.

Supply pipes: If you are lucky, the bendable connectors you have fitted will reach the severed pipe ends; if not, you must add straight sections of pipe joined with push-fit or compression fittings to bridge the gap.

With copper pipe:
• When measuring up for a new section don't forget to allow for the length of the fittings—not forgetting the 20mm of the pipe that gets 'lost' in the joint fitting.
• All pipe ends being joined must be cut absolutely square; if you find one that isn't, recut it rather than risk leaks later.
• Deburr cut pipe ends thoroughly—use a round file on the inside, a flat file on the outside. For a compression joint, chamfer the outer edge of the pipe end slightly.
• Smear jointing compound or wrap PTFE tape around compression joint threads before assembly, to guard against leaks and protect against overtightening.
• Assemble compression joints hand-tight

and then tighten one and a half turns further with your wrenches.
• Try not to bend a bendable connector more than once in the same place—it'll split. Avoid tortuous bends, too, especially in two places as connectors are easy to snap.

For plastic pipes:
• Always cut square and deburr.
• Whether you use compression or push-fit connectors, you need to strengthen the end of the plastic supply pipe with a metal insert.
• Smear silicone lubricant or washing up liquid on the pipe end before making a push-fit joint, otherwise the fitting may be damaged.

All pipe runs should be as unobtrusive as you can get them. On a pedestal basin, bend the connectors back towards the front of the basin and run the supply pipes to meet them, up the inside of the pedestal itself.

Drainage: Make up any gap between the new trap and the waste outlet with a new section of pipe. Use compression joints, rather than welded, as these will take up the slight variations in pipe size that occur between brands. Make sure the pipe ends are cut square and deburred; lubricate with

5 *Level the bath by adjusting the feet*

6 *Secure the bath to the wall by its frame brackets*

7 *Use a level to position the battens on the floor*

8 *Shape the bath panel, if necessary, using a saw*

9 *Secure the panel under the lip of the bath*

10 *Screw the basin or its brackets firmly to the wall*

washing up liquid before fitting.

It is quite conceivable that the old and new pipes are at different heights. On a bath, you ought to be able to get them level by readjusting the bath feet or even packing under the feet with scraps of plywood. Where a basin outlet passing through a wall is concerned, you can get various sizes of P traps, one half of which can be trimmed down to fit. Although you may have to cut back the existing waste pipes in order to

11 *On a pedestal basin, you can run the supply and drainage pipework unobtrusively behind the pedestal*

12 *When sealing around fittings, push the sealant ahead of you, gently squeezing as you go*

★ WATCH POINT ★

Fill the screw sockets with plumber's putty before tightening, then stop screwing in when the putty is almost all squeezed out—it stops the china cracking accidentally.

insert new sections, there should be no need to (and you mustn't) alter the fall to the stack or gully to any appreciable degree.

Installing the WC

With the pan temporarily in position, your first job is to measure and mark off what needs to be done to the existing soil outlet pipe. If it is too near or too far from the pan spigot, it must be cut to size.

Before you commit yourself, however, double check the relation of the pan as it stands to the cistern and the wall. On a close coupled suite in particular, there must be enough room to fit the cistern between the pan and wall, but not so much as to leave an unsightly gap behind the cistern once it is in its final position.

If you can re-use the soil pipe socket, chip out what remains of the old pan spigot and joint very carefully with hammer and cold chisel; keep going until it is completely clear of debris.

If the pipe is too short or too long, it must be cut using an angle grinder or cutting wheel attachment. Wear goggles to do this.

Angle grinders are dangerous tools: make sure that you grip the tool firmly and

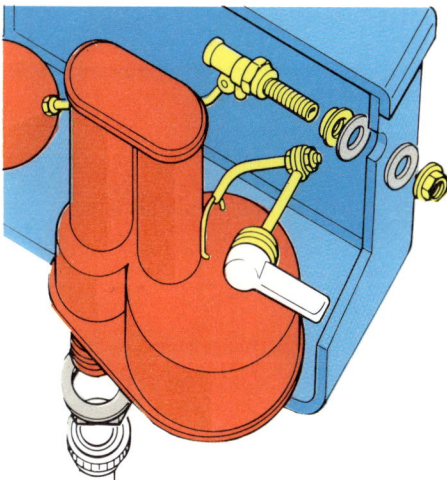

Assemble the inside of the cistern, watching out for the washers.

★ WATCH POINT ★

If the pipe is vitrified clay, wrap a piece of thick cardboard around the cutting line you have made and score against it, right around the pipe, using a tile cutter. The score will ensure a clean cut.

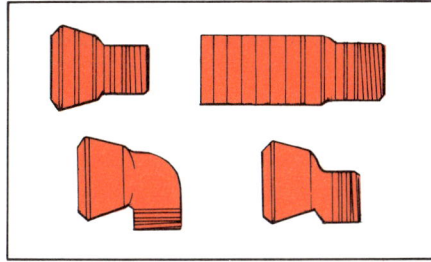

Above: you can connect the WC outlet to the soil pipe using a range of different adapters

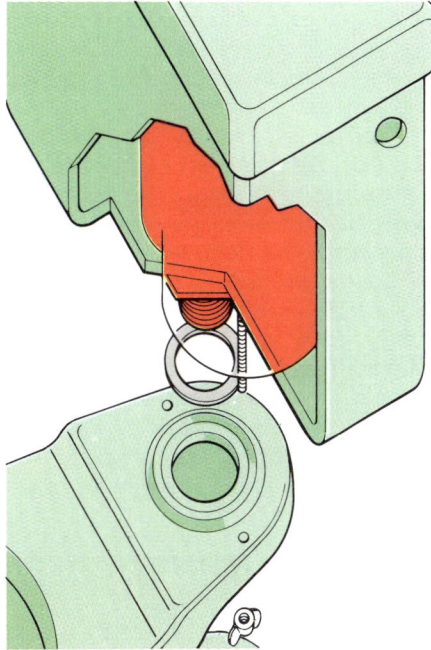

Bolt the cistern of a close coupled WC to the pan with wingnuts. Don't forget the washer

don't let it run away from you.

Once you have cut the pipe, slip the patent connector onto the end of the pan spigot and offer up the pan. (Do the same if you are reusing the socket.) Slip the other end of the connector onto (or into) the soil pipe and juggle the position of the pan until you get a perfect fit. Now mark the positions of the pan fixing holes.

Remove the pan to drill the holes (and plug them, if the floor is solid). Then replace the pan and connector, adjust as necessary, and screw the pan in place. Don't forget to fill the screw sockets with putty prior to tightening or the china may be cracked.

Now you can fit the cistern. On a close coupled suite, centre it on the pan, mark the wall fixing holes and remove. For a low level cistern, offer it up to the wall and get an assistant to help you centre it while you mark the fixing holes. Screw the cistern to

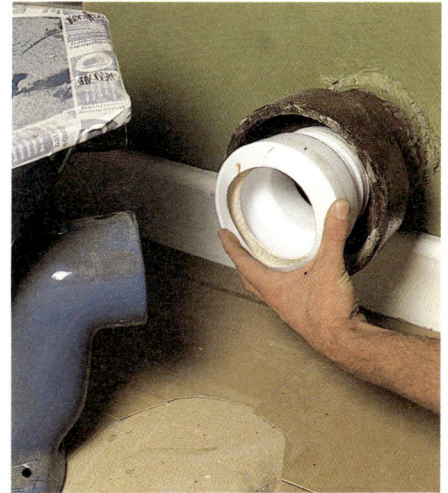

13 *Use an adapter to fit your new WC pan to the soil pipe*

14 *Secure the pan to the floor with brass or plated screws*

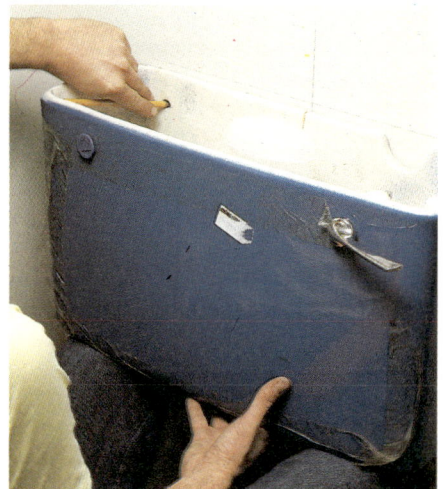

15 *Level the cistern and mark screw holes through it on the wall behind. Drill and fix*

16 *Use an adjustable spanner to tighten up the compression joint that connects the water supply to the WC cistern. Check for leaks*

17 *Follow the manufacturer's instructions when you assemble and fit the cistern, paying particular attention to the washer sequence*

the wall. Then, where necessary, screw the flush pipe to the siphon outlet, slip the other end into the pan, and fit the plastic sleeve that seals it.

All that remains is to connect the supply and run out an overflow. The connection for the water supply at the ball valve may be a compression joint or, more likely, a screw joint like the taps. The overflow should be 22mm copper or plastic. Drill the hole through the wall before assembling the run; seal it afterwards by forcing in some non-setting mastic from a gun and patch the hole on the outside wall with exterior filler.

Adapters for both feed and waste pipes will enable you to join modern Metric pipes to old Imperial ones

Connecting to old pipework

This is often the greatest plumbing problem in an old bathroom. On a job of this size it is usually easier to replace an old run in its entirety, rather than trying to extend it.

Supply pipes: If your supply pipes are exclusively lead, scrap them completely and replumb new runs in copper or plastic back as far as the cold storage tank. If they are only partially lead, they will have copper stubs wipe-jointed onto them—in which case connect to these in the normal way. Making your own wiped joints between copper and lead is difficult, and is best left to a professional. In any case, the lead pipes will have only a limited life.

If the supply pipes are galvanized iron, you should be able to buy adapters enabling you to tee-joint new runs in plastic pipe. But don't join iron to copper: the electrolytic action which this sets up will cause the pipes to decay prematurely.

It is quite likely that any existing copper supply pipes are Imperial rather than Metric. You can check—½in. and ¾in. Imperial pipe is measured across the internal diameter, Metric 15mm and 22mm across the external diameter—but in practice it can be quite tricky to be sure of the difference.

If you use a compression joint, you can join 15mm pipe to ½in. pipe without difficulty. To join 22mm to ¾in. requires a special olive. All capillary fittings require special Imperial-Metric adapters.

Drainage: If your waste pipes are lead, your house almost certainly has a two pipe drainage system. In this case, it should be fairly simple to fit a complete new waste pipe in plastic to a hopper head or gully.

If you are connecting into plastic pipes, take a sample with you to the plumber's merchant and get the brand matched—each brand has very slight size variations.

Adapters are available to connect modern 32mm and 38mm pipe to Imperial 1½in. and 1¾in. pipe and also to deal with other, more obscure, Metric sizes, such as 40mm.

If you are not sure of the brand but know the diameter and can match it, use compression fittings with a flexible rubber seal inside (rather than solvent welded joints) to take up any minor variations.

Finishing off

Once you have reinstated the water supply and checked for leaks, the new suite is ready for use. But you should make sealing around the bath and basin a top priority to protect the wall and floor behind.

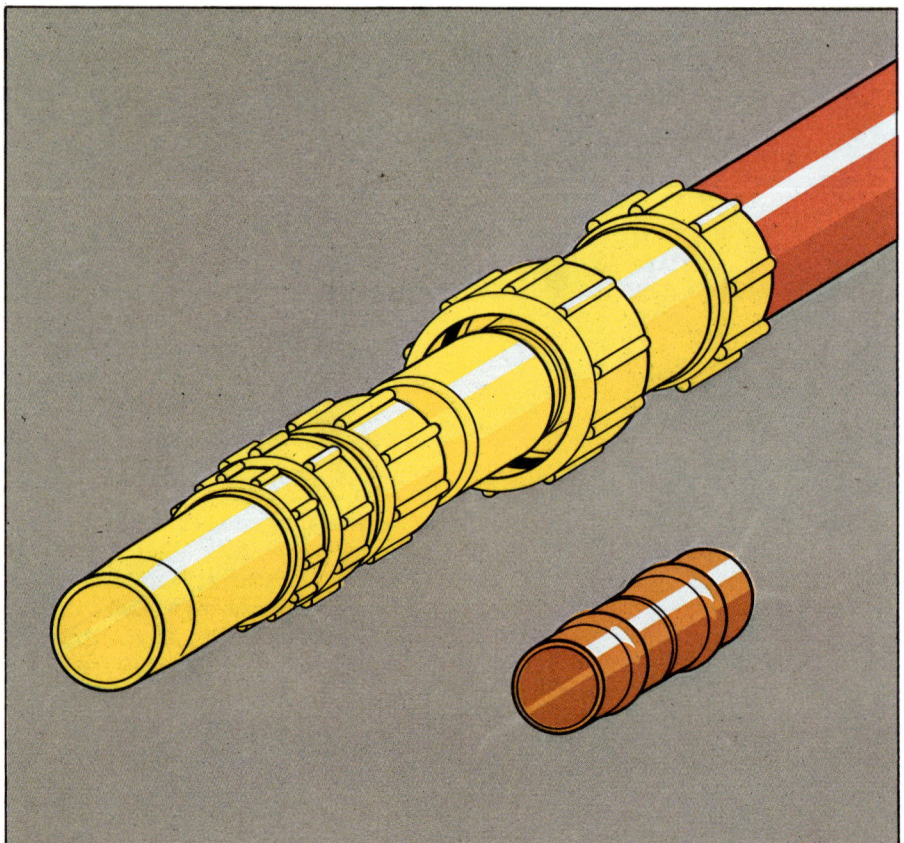

PLUMBED-IN SHOWER

Not only does a permanently plumbed in shower add significantly to the value of your home, but a shower is a hygienic and refreshing way to wash. Another convincing reason for having a shower is that they are economical to use—a five minute shower uses only a quarter of the water needed for the average bath.

Choosing a shower

The simplest type of shower is undoubtedly a rubber push-on hose which is little more than a glorified handspray. This type requires no plumbing but is clumsy to use.

A much better option is a **bath/shower mixer**. This replaces the existing bath taps and has two outlets for the water—one downwards through a single spout into the bath and one upwards to the shower. Changing from one to the other is simply a matter of moving a lever. Bath/shower mixers are sometimes sold complete with a wall bracket or rail for mounting above the bath. These allow the height of the shower to be varied to suit individuals.

Before buying a bath/shower mixer, make sure that it will fit—or can be adapted to fit—the tap holes in your bath.

A third alternative is a **shower-only mixer** which has its own independent hot and cold water supplies. This type gives a

better flow rate than a bath/shower mixer and can be mounted over a bath or in a separate shower cubicle. However, the snag is that, depending on where you put it, there may be a lot of extra plumbing to do.

Shower-only mixers can have one of

You can provide a good shower by fitting a shower-only valve (with blue rose) or a bath/shower mixer unit. Poor water pressure can be improved by fitting a spray booster (above left) or a booster pump (above right)

three types of control: either two knobs with one to regulate the temperature and one to regulate the flow; or hot and cold taps; or a single control to regulate temperature only.

You can get two designs of shower-only mixers—exposed and concealed. With an exposed design, the valve is surface mounted, together with some or all of the connecting pipework; with a concealed design, the pipework and valve are hidden.

Shower mixers have either a rigid or flexible hose which is mounted on the wall to either a bracket or rail.

The last option is an electric shower, which is easy to fit and connect but does not usually give the same flow that a fully plumbed in shower will guarantee.

Siting the shower

You have a choice of putting your new shower over the existing bath, or in a separate cubicle with its own shower tray.

The main advantage of installing a shower over a bath is that the plumbing will be simpler—you won't have to put in a separate waste pipe. On the other hand, if you're prepared to put in the extra pipe-work, there are several benefits to a separate shower cubicle—chiefly that you can put one almost anywhere in a bathroom or small cloakroom, for example.

Plumbing considerations

For bath/shower mixers and separate shower mixers, there are several requirements which will affect whether you can install them and, if you can, the amount of plumbing that you will have to do.

• The hot and cold supplies to the shower must be at the same pressure. With the majority of houses this won't be a problem since they have a cold water cistern which supplies both the cold taps—apart from the kitchen tap—and the hot water cylinder.

However, you may have a system where the hot water cylinder is supplied from a cistern, but all the cold taps are supplied from the rising main. With this type of plumbing, there's a good case for installing an instantaneous electric shower directly to the rising main.

The other type of plumbing system you may have is where all the fittings are connected to the rising main, the hot water being supplied by means of a multipoint gas heater. Again this may mean using an electric shower.

• The second requirement is that the pressure of water for non-electric showers must be sufficient to give a decent flow of water—5 litres per minute. The pressure is measured by the 'head' of water which is taken as the vertical distance between the shower rose and the bottom of the cold water cistern. The minimum head you need is 1m. If you have an insufficient head, it's best to install a shower pump which will boost the flow.

• The third consideration when fitting a shower is safety. If the cold supply to the shower is on the same pipe run as other fittings, the cold supply will be affected when the other fittings are used. This can mean the person using the shower could get scalded when the WC is flushed. The two ways of avoiding this are to provide a new and separate supply pipe for the shower from the cold water cistern; the other is to buy a thermostatically controlled shower mixer valve. With this type of mixer the outlet temperature is controlled so that if

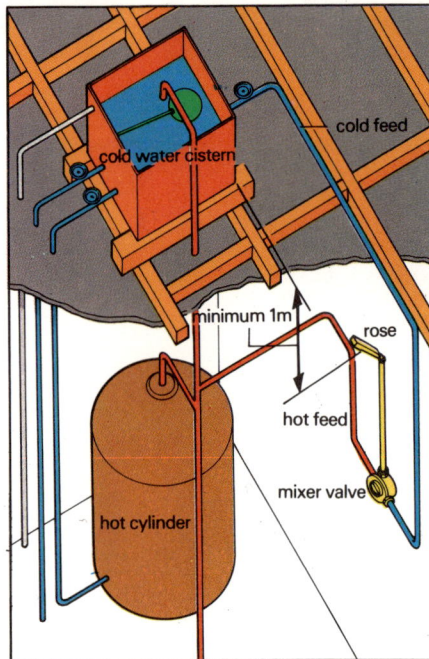

Water supplies can be taken from the cold cistern and the hot cylinder. You need at least 1m of 'head' above the rose for a good shower

the pressure on the cold side drops, the hot flow is reduced accordingly.

The reverse situation can occur with the hot supply—resulting in the shower running cold when a hot tap is turned on somewhere else in the house. This is annoying rather than dangerous, but can be avoided by making sure that the hot connection to the shower is the first from the pipe leading out of the top of the hot water cylinder, before the take-off to any hot taps.

Additional materials and tools

Apart from the shower fitting itself, which should come complete with instructions and all the nuts and washers you need to install it, you may need certain extra materials and fittings.

For a bath/shower mixer, you will certainly want PTFE tape and possibly flexible push-fit tap connectors. For a shower-only mixer, you will want lengths of 15mm flexible piping plus sufficient elbows and T-connectors to complete the supply runs. You may also use wall plugs, pipe clips and either plaster or filler to make good the installation.

If you decide to incorporate a booster pump into the pipework, you have a choice of fitting two types. One goes between the

mixer and the shower spray, while the other is fitted before the mixer to both the hot and cold supply pipes.

To save yourself a lot of extra wiring, it's best to buy a pump which switches on automatically when the shower is used—otherwise you will have to install a ceiling mounted pull-cord switch. It is possible to buy some pumps which have a built-in transformer and operate off low voltages—either 12V or 24V. With these, switches can be wall-mounted and within reach of the shower.

For a bath/shower mixer, the one essential tool is a bath/basin spanner, and you will probably need an adjustable spanner as well, to tighten up the tap connectors. You'll need several tools for a shower-only mixer including a drill and bits, saws and screwdriver.

★ WATCH POINT ★

If neither of the supply pipes has a gate valve, drain the cold water system by tying up the ballcock in the cold water cistern and opening both bath taps. Don't turn off the main stopcock or your kitchen tap will be without water.

Fitting a bath/shower mixer

Installing a bath/shower mixer is a straightforward job, as there is no extra plumbing to do. Fitting the mixer is very much like installing any other sort of tap. Before removing the old taps, close the gatevalve on the cold water feed from the cistern and drain down the small amount left in the pipe by opening the cold bath tap. Similarly, turn off the water heating and close the valve supplying the hot water cylinder.

Getting at the bath taps can be awkward—there's not much space at the end of a bath and you'll have to remove the bath side panel to give yourself access.

Disconnect the tap connectors with an adjustable spanner and then use a bath/basin spanner to undo the backnuts holding the taps in place. Be careful not to damage the bath and, if necessary, get someone to hold the taps so that they don't swivel around as you loosen the nuts.

When you've got the old taps off, clean any dirt or old compound from around the holes and polish the ends of the supply pipes with wire wool.

Before attaching the new fitting to the

bath, push on the gasket which goes between the bath and the mixer.

If you need to move the positions of the tap connectors because the shanks on the mixer are either too long or too short, cut back the supply pipes and fit lengths of flexible copper pipe which have tap connectors at one end.

The mixer is fixed to the bath by two large backnuts—these usually need plastic washers and sometimes metal ones, too, between them and the bath. Put the plastic washers on first, followed by the metal washer and then the nut. Tighten up the nuts with a basin spanner.

Before screwing on the tap connectors, wrap a strip of PTFE tape three times around the threads on the shanks of the mixer to seal the joint.

Fix the hose by simply screwing it to the top of the mixer valve—again using PTFE tape to get a watertight join.

If you have one, fit the wall bracket or sliding rail next. Measure the optimum height for the rail—or bracket—and mark its position on the wall. Check that the rail is vertical with a spirit level. Drill clearance holes for suitable wallplugs and fix the fitting in place with chromium plated screws to avoid rusting.

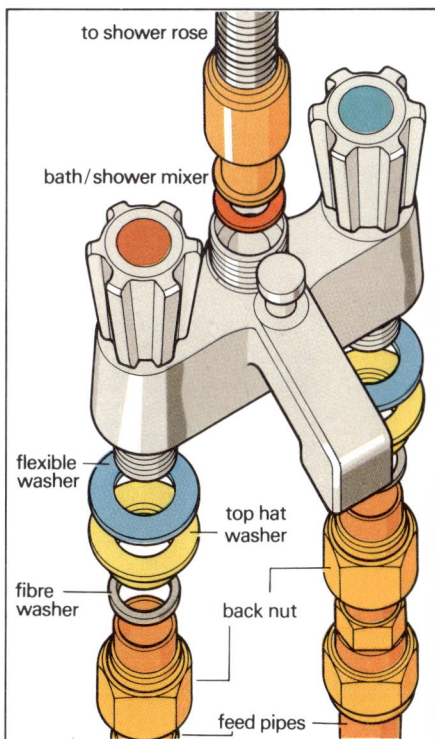

A bath/shower mixer (right) can replace existing taps if the hole separation is suitable. The feed pipes are connected in the same way with a back nut tightened onto the tap tails

After fixing a shower over a bath, you'll need some kind of shower curtain or screen to protect the surrounding idea.

1 *The first step in replacing pillar taps with a bath/shower mixer is to loosen the tap connectors*

2 *With the tap connectors disconnected, use a bath spanner (crowsfoot) to undo the backnuts holding the taps*

3 *Position the mixer unit, and tighten the back nuts and tap connectors*

4 *Screw the shower rose support bracket to the wall, slip on the chromed shroud and press the rose into place*

Connecting to the supply

New cold water supply: First turn off the water supply to the cistern or, alternatively, tie up the ballcock. Drain the whole system by turning on the bathroom cold taps.

Near the base of the tank, mark the position of the new hole. If you haven't got a tank cutter or hole saw, drill a series of small holes inside a 22mm circle and file the edges clean.

The new pipe is connected to the cistern with a 22mm tank connector. Make sure that you put a nylon washer on either side of the cistern before tightening up the flanged nut on the outside.

Cut, and then fit, a 150mm length of 15mm pipe to the compression end of the connector.

To the other end of the short pipe, fit an isolating wheel valve—again a compression fitting—and from this run 15mm piping to the cold inlet of the mixer.

Tee-ing into an existing pipe: After draining down the system, mark on the pipe where you want to put the fitting—remember that this should be closer to the hot cylinder or cold cistern than any other of the branches.

The gap you need to cut in the pipe to take a tee fitting is somewhat less than the length of the tee—to allow for the length of the pipe that slots inside.

Mark the gap on the pipe with a pencil and cut at right angles.

Insert the 22mm × 22mm × 15mm reducing tee into the gap and tighten up the securing nuts on each end of the tee.

Connect the 15mm pipe to the reducing tee and run this to the appropriate inlet on the mixer valve.

INSTALLING A SHOWER TRAY

A shower is economical and quick to use, more convenient—and less wasteful of water—than taking a bath; it also occupies very little space. If you want the shower to be a completely independent fitting, you must install some type of shower tray, an enclosure or splashback, and a new waste drainage system.

Providing you've got the space you can still install your self-contained shower unit in the bathroom. But if you can arrange the drainage there's no reason why you can't fit it elsewhere—in the bedroom, inside a fitted cupboard, or even in a room of its own—and add considerably to the value and luxury of your home in the process.

Planning considerations

Where you position your shower depends largely on how you're going to dispose of the waste water. Most modern waste systems have the 'single stack' arrangement: all soil and waste pipes discharge directly into a large pipe connected to the drain. But in many older houses the soil waste is dealt with by a separate pipe. The baths and basins upstairs discharge into an open hopper head at the top of another pipe, or downstairs into a gully. Both of these lead in turn to the same drain. In fact, whether you have a one or a two-pipe system, it's usual for sinks and basins on the ground floor to discharge into a convenient gully.

If your shower is to be on the ground floor it's simplest to dispose of the waste water direct into a gully; if it's on an upstairs floor you can either take the new waste pipe to an existing hopper or connect it into the combined waste/soil stack.

Depending on the age of your house, the stack may be made of cast iron or PVC. Leading a waste pipe into a PVC stack is straightforward, although you must identify the make and buy compatible connection fittings—systems aren't usually interchangeable. Cast iron stacks cause more problems: you'll need special connection fittings for these.

Check with your local authority on the recommended pipe fittings—they may specify certain types. Some authorities don't accept push-fit connections and may demand that the first 2m of soil pipe is cast iron.

Once you've worked out how you're going to deal with the waste you can plan your pipe run. This shouldn't exceed a length of three metres for efficient drainage and it should have a slope downwards of between 1° and 5°—equivalent to a drop of between 18mm and 90mm per metre length. Avoid too steep a slope; this can siphon water from the waste trap under the shower tray and such a seal is necessary to prevent smells from entering the house.

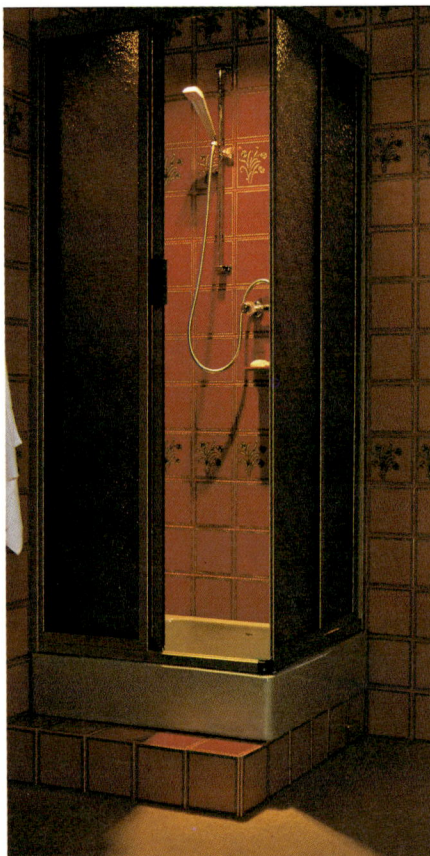

A shower tray and surround can be fitted almost anywhere. If you want to build your own plinth, see page 39

Choosing and siting a tray

Shower trays are commonly made in acrylic plastic or glazed ceramics, although you can also buy pressed steel or cast iron types. Acrylic trays are light in weight and come in a variety of colours to match standard ranges of sanitary ware. There's a range of shapes and sizes—triangular, quadrant and square are typical—to suit most locations.

Most types are set on legs and have detachable side panels, giving space underneath for the outlet connections, and some have a raised flange on two sides for fixing the tray direct to the masonry before you start tiling over.

Ceramic trays are also made in various designs and colours, although they don't usually incorporate space underneath to accommodate the trap. Also, they're very heavy and you'll need help to fit one.

The waste outlet—like an ordinary plughole but without the plug—may be in the centre of the tray, at one side, or in one corner, so you can position the tray exactly where you want.

The trap should be the shallow 38mm diameter shower/bath type, with a 70mm seal for single stack drainage or a 50mm seal for two-pipe drainage. Use a P-trap where the waste pipe exits direct to the outside, or an S-trap where the pipe must include an internal drop. If your chosen tray doesn't include space underneath for the trap you'll have to cut a hole in the floor—if it's a timber one—and run the trap and waste pipe in the space below. Alternatively

you can mount the tray on a plinth (see Constructing the plinth) to accommodate the trap and pipe. Include an inspection panel for access.

Don't lead the pipe across the joists: you'd have to cut a notch about 40mm deep in the timber, and this would seriously weaken it.

Don't forget to allow sufficient space outside the shower for drying off: you'll probably need about 700mm on the opening side of the tray.

The best place for your shower tray is in a corner—the walls can form part of the enclosure, as long as they're properly waterproofed. For this you can either tile the

surface or fix sheet plastic, glass, or waterproof decorative wallboard.

Ready-made enclosures are available in kit form. They usually consist of an aluminium framework with plastic or glass panels, which you attach to the wall on top of the tray. Most types are adjustable to fit various tray sizes, and some incorporate a system to align the enclosures on out-of-true walls. They generally have sliding or concertina-type doors, which you can also buy separately for fitting to a home-made surround. Various designs of enclosure are made for the different types: freestanding, corner or built-in showers.

It's important to make sure that the

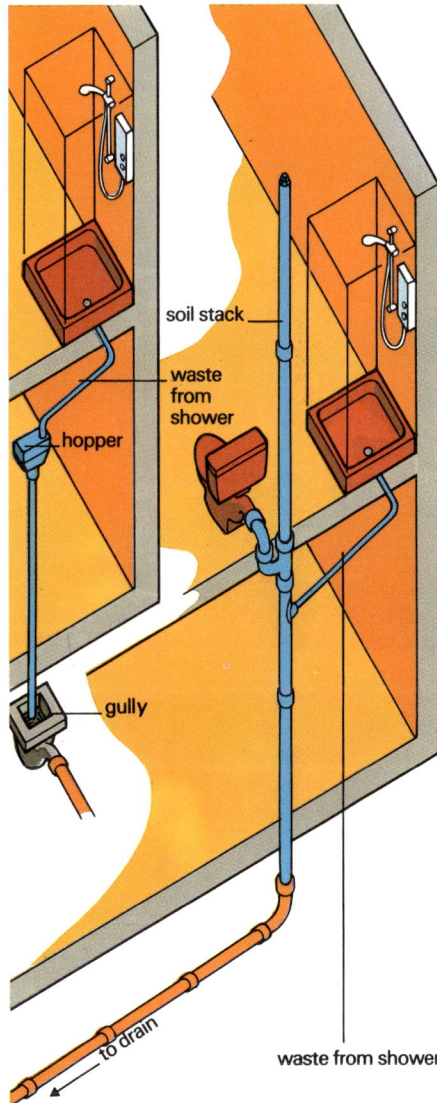

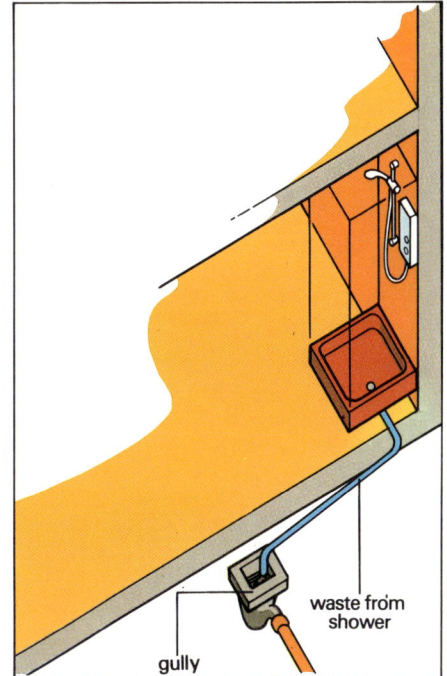

Plan your shower tray's waste disposal. In most modern houses, all soil and waste pipes discharge into one soil stack; in older houses, the shower will need to be fixed up so that it discharges into a gully via an open hopper head

Regardless of the age of the house, it is usual for sinks, basins and showers on the ground floor to discharge directly into a gully. This is usually for simplicity's sake—connecting into a soil stack can be tricky unless there is a convenient hopper head to use

joints between the wall and tray and the surround and tray are sealed against water seepage, using quadrant tiles or a flexible non-setting mastic to fill the gap, followed by an acrylic or silicone sealant. If you fit a shower surround, this too must be sealed so that water cannot get in (see page 44).

Fitting the shower tray

Once you've planned the route of your shower waste pipe you can install the tray in

1 *Put the tray in position and adjust the legs until it sits level*

2 *If your tray is fitted with wall brackets, mark their position*

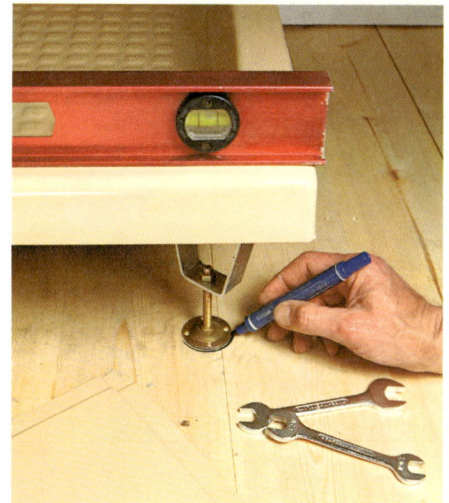

3 *With the tray's position finalized, mark around the feet with a pen*

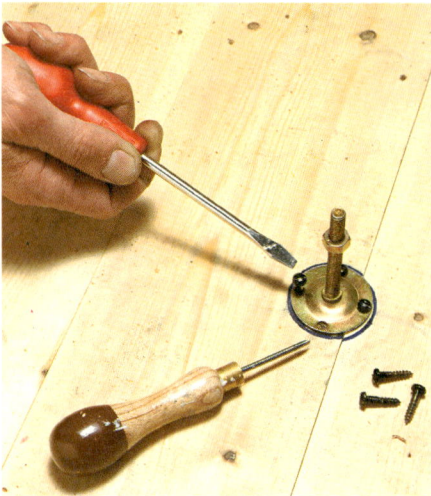

4 *Remove the feet from the tray and secure them to the floor*

5 *Relocate the tray on its feet, add the securing screws and check for level*

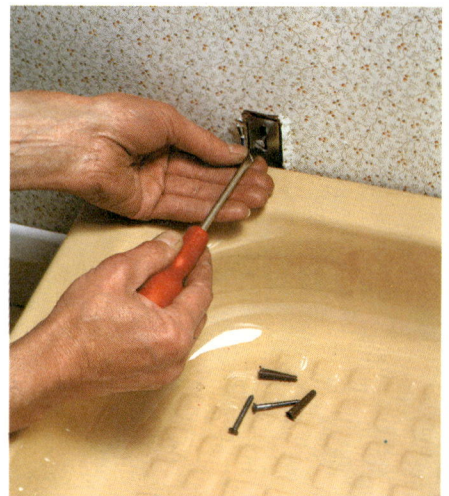

6 *Screw the tray's fixing brackets to the wall. You can make good later*

7 *When you have made the waste connections, fit the side panels*

its final position and attach the trap.

If you're installing an acrylic shower tray with its own plinth all you have to do is fit the waste outlet and trap, attach its short legs and set it on the floor in the position which you require.

The type of supporting leg varies from make to make but they usually consist of angled metal brackets which screw or clip onto the underside of the tray, with threaded feet that can be adjusted for level. Attach the legs first.

Place the tray against the wall. Mark the area of skirting board that needs to be removed so that the tray can be placed flush with the wall. Remove the tray then mark the skirting for cutting. Prise away the skirting board with a claw hammer or bolster chisel to give yourself room to cut

down the lines with a small saw. Protect the wall by levering against a scrap of wood.

Reposition the tray against the wall and adjust the nuts on its feet so that it sits level on the floor; check with a spirit level. If your tray includes a wall bracket, mark its position on the plaster. Hack off a band of plaster so that the tray can be fixed to the

★ **WATCH POINT** ★

It's a good idea at this stage to temporarily assemble the waste connections to your stack or gully, and make any adjustments that may be necessary to avoid straining the pipes or connections.

masonry and the bracket plastered over. Mark around the position of the feet on the floor with a pen.

Turn the tray upside down and remove the feet. Place them over your marks and fix them to the floor with screws (and plugs on a solid floor). Drill and plug the wall to take the wall brackets. Carefully locate the tray on its feet and add the securing nuts. Screw the brackets to the wall.

Reconnect the waste fittings and pour some water into the tray to check for leaks. Separate side panels usually clip onto the tray and locate on the floor battens. Fix these when you've finished making all the waste connections.

Ceramic shower trays sometimes have separate outlet inserts. To fit one of these, press a sausage of plumber's putty around the underside of the fitting then insert the outlet into the hole in the tray and fit its metal washer. Wind PTFE sealing tape around its thread. Screw on the retaining backnut and tighten with an adjustable wrench. This will squeeze putty from the perimeter of the outlet, which you should remove immediately for a neat finish.

Making a water-tight seal around the outlet is one area where many DIY enthusiasts come to grief—but there is no reason why they should. Apply plenty of plumber's putty to the fitting and ensure that it is evenly distributed. Be thorough in your use of the PTFE tape too and then, when you tighten the backnut, you should have a good seal.

On trays that have no built-in plinth you'll have to construct your own (see Constructing a plinth). This is not difficult but you must give careful thought to the position of the waste outlet and any other features peculiar to your room alone.

Cutting a hole for a pipe

When cutting a hole for a waste pipe, you cannot afford to adopt a hit or miss attitude—so draw up a plan.

Basically, you need to construct a square on the interior wall using a reference point common to both sides of the wall—a window for example—and then, using the measurements of the square, draw an identical square on the exterior. Then knock the hole through from both sides, so that you create a neat opening in the middle of the wall.

To cut a hole in your wall to take the waste pipe, first mark its position. Do this by temporarily attaching a short length of pipe to the waste trap so that it touches the

Using a 15mm masonry bit, drill a series of holes around the waste pipe guidelines

Knock through the wall with a club hammer and chisel. Work from both sides

Test fit the waste pipe in the hole— remove it when you're satisfied

wall. Mark the wall neatly around the pipe.

On a thick cavity or 225mm wall it's best to mark the position of the hole inside and out, and then tackle the cutting from both sides. To do this you'll need a common reference point on both sides of the wall, such as a window or another pipe passing through nearby. Measure the height of the reference point from inside floor level and transfer this measurement to the outside. You now have the floor level marked on the outside wall.

Using a spirit level draw a vertical line from the floor level inside to your reference point. You can then transfer this line to the outside wall.

Draw a second vertical line from the inside floor level to pass through your pipe exit point. Measure the distance between the two vertical lines and transfer this line to the outside wall. Inside, measure from floor level to the centre of the waste pipe exit point and transfer this measurement to the outside wall. This gives you a reference for marking the pipe exit point on the outside of the wall.

To cut the hole, starting at one side of the exit point use a 15mm masonry drill to make a series of holes around the perimeter of the guideline. Hack off the plaster from within the guidelines using a slim cold chisel and club hammer. Start to cut out the masonry from the hole, then move to the other side of the wall and repeat the procedure until you break through. Try to keep the sides of the hole as square as possible.

If you're working up a ladder when knocking through from the outside, keep a lookout for people passing below you or cordon off the area.

Constructing the shower plinth

If your shower tray has no integral base with room to take the waste trap, you can make a timber plinth and tile it to match the shower surround or splashback. In this way the plinth is completely integrated.

The plinth is made from four pieces of 150mm × 50mm PAR softwood and one piece of 100mm × 50mm softwood held together with proprietary joint blocks or battens that you can make yourself.

Its overall height—150mm—leaves ample room for the shallow waste trap and there's a gap in the frame to provide access in case of blockages or leaks; you can fit a trap door or leave the access point open.

First take the measurements of the base

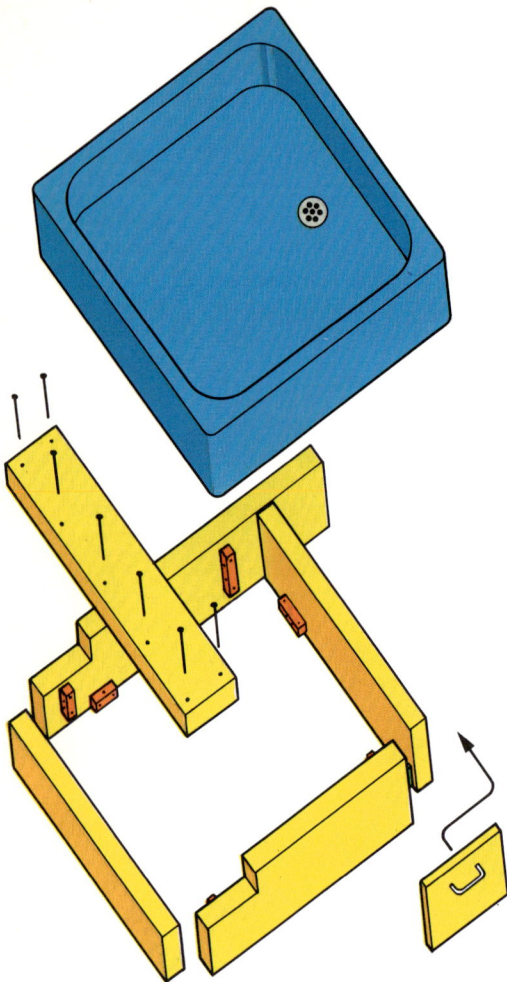

A basic plinth design. Note the access panel and the side panels which are notched for the step

of your shower tray. Turn the tray upside down and measure the distance from the edge of the tray to the inside edge of the outlet. Add on about 50mm for clearance. From this point measure to the front of the tray. Add on 150mm for the step.

Mark out and saw two pieces of timber to this length to form the sides of the plinth. Use a try square when doing this.

Measure the width of the shower tray and subtract 100mm—twice the timber thickness. Cut one piece of 150mm × 50mm and the piece of 100mm × 50mm to this length, to form the front and back of the plinth.

So that the step can be set flush with the top of the plinth, cut a notch in the front edge of both side pieces 150mm wide by 50mm deep.

Assemble the frame with two joint blocks per corner to form a square, with the two notches at the top front edge. As the plinth is to be a permanent fixture you should use the one-piece joint blocks instead of the two-part knock-down type.

1 *Fit a separate outlet insert with plumber's putty*

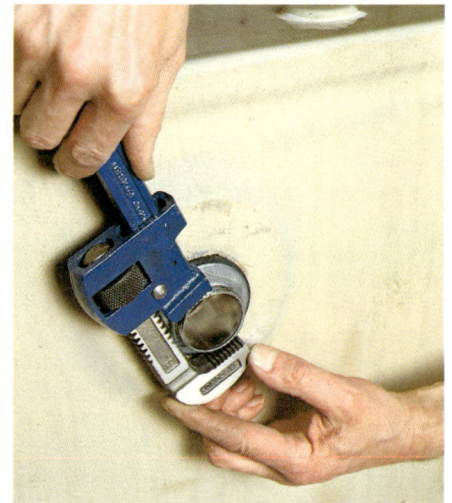

2 *Screw on the retaining back nut and tighten it with an adjustable wrench*

3 *Use the tray as a guide to mark the sides of the plinth for cutting.*

4 *Lay the sides in position and use them to mark the front and back for cutting*

5 *Cut a notch in the top front edge of each side piece ready for the tread*

6 *Use joint blocks or battens to assemble the plinth framework*

7 *Lay the trap on top of the plinth. Adjust until the two are aligned as accurately as possible*

8 *Secure the plinth framework to the floor with joint blocks or battens*

9 *Cut and fit the tread and riser for the plinth step. Hammer all nails home fully for a secure fixing*

Lay the assembled frame on the floor and place the shower tray on top. Mark the position of the plinth and then remove the tray. Place joint blocks inside the plinth and screw them to the floor and frame.

Return the tray to the plinth and make the waste connections so they are tight and free from any possible obstruction.

Cut a piece of 150mm × 50mm timber to fit across the plinth and attach it to the notched section with waterproof adhesive and 75mm long nails to form the step.

Finish off the plinth with paint or by cladding it with ceramic tiles to match the walls or splashback. If you tile it's best to recess the plinth by the tile thickness so they're exactly flush with the sides of the shower tray.

Arranging drainage

There are various drainage arrangements you can have for your new shower, depending on its location (see diagram right). If the shower is on the ground floor, simply lead the pipe from the waste trap through the outside wall into an existing gully; if the shower is upstairs you can lead the waste pipe into a hopper head or make a separate connection into the soil stack. Each of these leads in turn to the same drain. If you use plastic solvent-weld or push-fit plumbing connections, the job is very straightforward and requires no special tools or equipment.

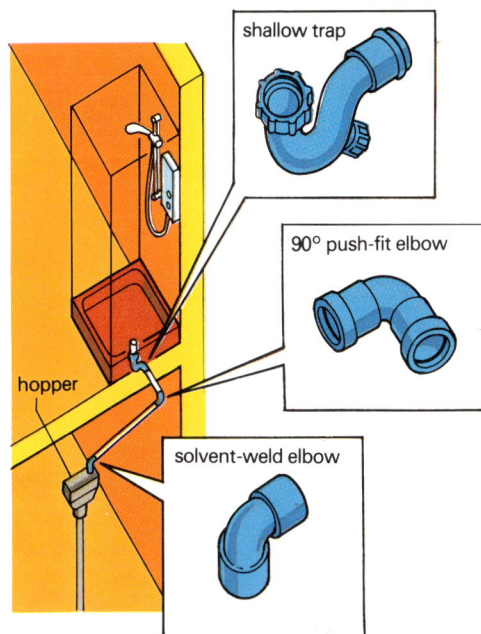

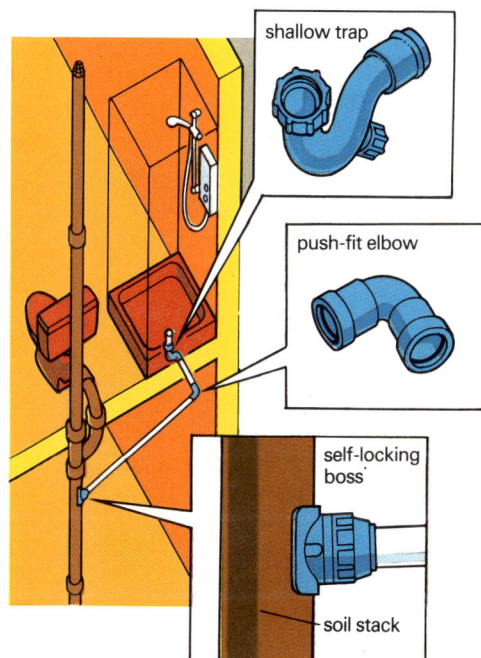

shallow trap

90° push-fit elbow

hopper

solvent-weld elbow

shallow trap

push-fit elbow

self-locking boss

soil stack

The waste connections for an independent shower are easy to make if you use plastic push-fit and solvent-weld fittings. You need few components to lead the waste to a hopper or gully (above left); even the soil stack connection needs only one extra component. The pipework is all made in 38mm plastic—so be sure to buy fittings compatible with pipes this size

There are, however, some special requirements to bear in mind when you're planning your waste run from the shower:
• for efficient drainage, the waste pipe shouldn't exceed a length of three metres;
• the pipe should slope downwards between

1° and 5° (a drop of between 18mm and 90mm per metre length);

•the waste trap should be the 75mm shallow type made especially for showers and baths;

•the waste pipe must not enter the soil stack within 200mm of a soil branch pipe connecting to the WC (unless you fit a specially designed deflector);

•don't take pipe runs across joists—the size of notch you would have to cut would cause a serious weakness in the timber.

Fitting a shower enclosure

There are many types of shower enclosure you can fit. They're available in kit form and fitted with patterned glass or acrylic panels in various colours.

You can buy various designs of shower surround depending on the location of your shower. These are commonly the corner surround, which consists of two panels (the walls act as the other sides); the built-in type (only one panel fits over a fitted unit) and the freestanding surround, which consists of three panels (the wall makes the fourth side). Some special panels include niches—often on both sides of the panel—to hold soap, towels, cosmetics and other bathing accessories but these tend to be expensive. You can usually fit the door in whichever panel you want for convenience, or even in two sides if there's sufficient room outside for drying off. Doors can be operated in a variety of combinations, usually concertina and sliding.

Before deciding on any particular type of surround, pay a visit to one of the large bathroom or DIY retailers to see what there is on offer. Many of the surrounds will actually be on full display in the store.

Once you've fitted the tray you can make up the waste pipe connections and erect the surround to complete the job.

Making the waste connections

When you've secured the shower tray to the floor, you can make the waste pipe connections into the soil stack, gully or hopper according to your domestic arrangements.

Once you've positioned the tray, cut an exit hole in the wall to take the pipe (see Cutting a hole for a pipe, page 39). To connect the tray to the trap, pass a length of 38mm plastic pipe through the hole from

1 *Once you've cut a hole through the outside wall to take the waste pipe, fit the trap to the pipe and test the fit against the tray*

2 *Outside, if you're on the first floor, cut the waste pipe to allow for a 90° elbow to be fitted to divert the waste to a convenient hopper*

3 *Attach a 90° elbow (here, a push-fit) to the protruding pipe then fit a short length of pipe to lead towards the hopper*

4 *Fit a 45° elbow on the end of the pipe to discharge into the hopper. Ensure that the pipe doesn't slope too steeply*

5 *Secure the pipe run at intervals with special plastic brackets which simply clip over the pipe and fix to the wall*

6 *To connect into a plastic soil stack, cut a hole in the stack, and stick in the entry piece of a self-locking boss*

7 *Attach the screw-on section of the boss, the collar, the rubber grommet and the push-fit connection*

8 *Cut a length of waste pipe to lead from the boss connection to the waste outlet, allowing for a 90° elbow conection*

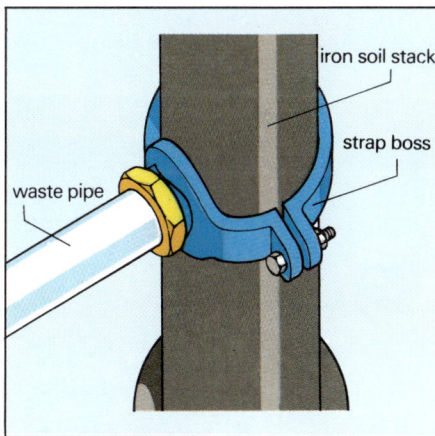

Connecting a waste pipe into a cast-iron soil stack is fairly straightforward, but you'll need a special strap boss connection

the outside. Connect it to the trap, making sure it's fully seated. Mark the pipe on the outside of the wall. You will need to cut it at this point so that you can fit a 90° elbow fitting on its end. This will divert

the pipe so that it runs to the drainage point.

Cut the pipe and attach the elbow using solvent-weld connections, or push-fit sockets. If you use a solvent weld fitting make sure it includes an access plug in case of any blockages.

Connecting to a hopper or gully: Measure how much pipe you'll need to join up the elbow to the hopper or gully, and cut a length to fit. If the pipe is leading to a grating over the gully you'll probably also need to fit a 135° elbow above the grating so that you can attach a pipe from this to below grating level. Make sure that the pipe through the grating stops short of the water level in the gully. You'll probably need to buy a new plastic grating and cut it to take the extra pipe.

If the pipe is going to a hopper, fit a 45° elbow where it discharges into the hopper head itself.

Clip the pipes in place with proprietary brackets at the required points—roughly 500mm intervals is ideal—and ensure that the pipes are aligned without any strain being exerted to hold them.

Connecting to a plastic soil stack: There are basically two methods of connecting your waste pipe into a plastic soil stack. You can lead the pipe to a convenient 'boss branch', a section of the stack which includes a spare entry point, or you can fit a special self-locking boss connection.

To break into a boss branch you have to cut out the circle of plastic within the entry point and fit a special socket adaptor. The procedure may vary slightly according to the make of pipe but usually consists of solvent-weld fittings.

Cut the hole in the pipe using a special 38mm diameter hole cutter attachment to an electric drill. If you haven't one of these you can drill a row of holes at the perimeter of the entry point in the boss and insert a padsaw blade to cut out the section. Be sure that the circle of plastic doesn't accidentally fall into the soil stack.

To fit a self-locking boss, cut a hole to take the entry piccc and attach the collar piece and pipe entry section. Fittings vary from make to make but usually consist of solvent-weld connections and adapters.

Connecting to a cast iron soil stack:

There are similar fittings for connecting up a plastic waste system to a cast iron stack. But most involve either cutting a hole in the stack to take the pipe and adapter, or the fitting of a special conversion socket or 'strap boss', which is clamped around the cast iron stack.

Cast iron isn't easy to cut without special tools so the easiest solution to the problem is to remove an entire section of the stack between the two joints and replace it with a plastic section. For this you'll have to fit an adapter at both ends of the run to take the new pipe. You can then fit a new boss branch in the run and connect your shower waste to it, as previously described.

Erecting the shower surround

Proprietary shower surround kits are available to suit most types of tray. The type you choose depends on whether you want a completely freestanding cubicle or just a corner arrangement.

There are many designs of shower enclosure but your choice will be governed, to some extent, by the position of your tray —in a corner arrangement, freestanding, or built-in. The basic corner arrangement consists of two panels—one or both including doors—which are each fixed to a wall. The walls make up the other two sides of the enclosure and then the shower head can be fitted on to whichever wall suits you better.

Freestanding surrounds consist of three panels (with various door arrangements) fixed against a single wall.

The third basic arrangement is used where the shower is fitted into an existing cupboard or alcove and is simply a single panel with door which completes an enclosure already three parts constructed.

When buying an enclosure, bear in mind that the panels containing the doors (sliding or folding) should be fitted in the best position for you simply to step outside where you can dry off.

Methods of assembling kit surrounds vary considerably from make to make but most consist of an aluminium frame with glass or plastic panels. Many include a feature that enables the frame pieces to be adjusted to suit different sizes of tray, and have easily assembled slot-in corner brackets—often made of plastic.

Fitting a surround usually involves the following procedure. Mark the position of the wall uprights on the wall using a spirit level to ensure they're truly vertical.

1 *Complete your shower tray plinth, once you've made the waste connections, with a tiled panel providing access to the trap*

2 *To erect the surround, place the first upright against the wall, check its level with a spirit level and mark the fixing positions*

3 *Drill and plug the wall in the marked fixing positions, return the upright to the wall and screw it in place. Take care not to damage wall tiles*

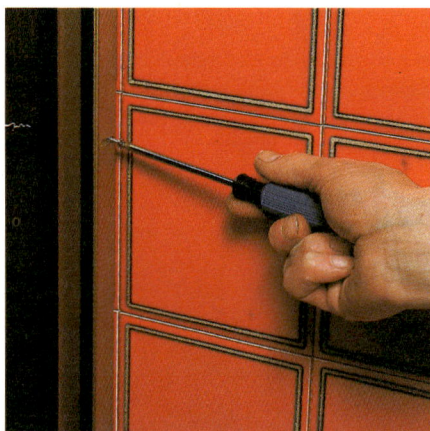

4 *On a flat surface, first assemble the top and bottom frames with the plastic corner brackets and then attach the glazed panels*

5 *Lift up the frame and panel assembly onto the tray making sure it's perfectly square and level. Adjust the screws to alter the level*

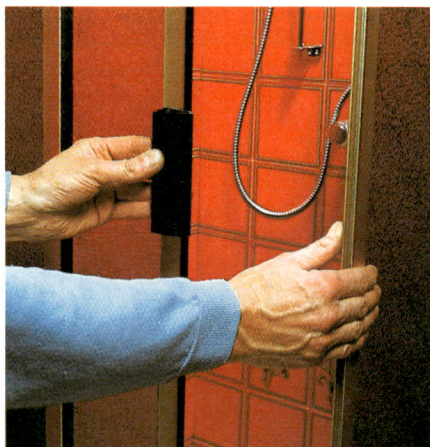

6 *Install the door panels in their tracks, fit the screw-on handle and test the operation of the door. Make any adjustments which are necessary*

7 *Carefully seal the join between the shower surround and the tray, and the surround and wall, using a flexible mastic sealer*

Attach the uprights with screws—usually provided in the kits—driven into wallplugs.

If you're fixing the surround to a tiled wall, stick a piece of masking tape to the tiles below the fixing hole positions. This enables you to mark the surface more easily and will prevent the drill bit from sliding on the glazed finish.

Attach the first side panel—which might contain the door—to the upright and adjust the length, if necessary, to fit the tray. Add the second panel and adjust its size. Locate the top and bottom runners in the corner brackets and again check that the frame is perfectly square. If the tiles haven't been fixed truly vertically you should adjust the level of the surround uprights to follow the line of the grouting. Otherwise the surround will appear crooked—even if it's not.

If you're making a corner cubicle you can then locate the door on its runners and check that it opens smoothly. If you're making a completely freestanding cubicle you'll have to add the front panel. Again check the operation of the door.

Should the door not slide smoothly you can lubricate the runners with a little vaseline although you should also check that there isn't any distortion of the frame.

Rubber seals in the frame may require lubricating with silicone lubricant—although you can use washing-up liquid—before you insert their panels.

To complete the cubicle you'll have to seal the joins between frame and wall, and frame and tray, with a proprietary sealant. Use a flexible waterproof sealant for this, taking care to apply it in one continuous bead along the full length of each join. Be particularly careful around the plinth area as that will be the most likely source of trouble.

TOWEL RAIL

A heater not only gives a special touch of comfort to a bathroom—ensuring warm, dry towels—but also helps prevent condensation. Many different types of bathroom heater are on the market, so you should be able to find one to suit your home.

Types available

Which type of bathroom heater you should choose depends a lot on your current heating arrangements and on the size of your family, so consider the various options before you start work. Listed below are the most common heating systems.

•**Wet central heating:** If you have a 'wet central heating system'—one with radiators—then arranging a bathroom towel heater is usually quite simple. You may already have a radiator in the bathroom—in which case all you need to keep towels warm and dry is a clip-on towel rail. These are relatively cheap, come in different styles and

sizes to fit almost any make and size of central heating radiator, and are very easy to fit: they just clip round the radiator without the need for any extra plumbing.

If you have no heater in the bathroom, you might like to add one. This is usually a relatively simple plumbing job consisting of connecting a couple of branches of new pipework to nearby central heating pipe runs. In this case, there are three types of heater you could install. The first is an **ordinary radiator**, probably with a clip-on towel rail. This is often the most sensible solution—radiators are relatively cheap, and because they come in a wide range of sizes you can be sure of getting one large enough to heat the bathroom thoroughly. The second type is a simple **heated towel rail**. This is basically a number of chrome-plated pipes joined together to form a towel rail. They give lots of space to hang towels on, but generally give off little heat; they can also be expensive. A **radiator/rail** is a combination of the two other types—a towel rail mounted on (or alongside) an ordinary radiator.

Whether you have a radiator in the bathroom or not, you might prefer to connect a heater into the hot water circuit—there are several definite advantages for this method depending on your plumbing system.

•**No central heating or hot water boiler:** Some houses, especially in countries other than the UK, do not have water boilers—heating may be provided by warm air units or electric storage heaters; hot water by immersion heaters or instantaneous boilers. Here, the usual answer to keeping towels in the bathroom warm is to install an electrically-operated oil-filled towel rail—see page 55. These are easy to install, but in most countries, electricity is an expensive form of heating.

•**Hot water boiler:** Some houses without a central heating system heat the hot water with a boiler of some sort which feeds a hot water cylinder. You should tap into the pipes connecting the boiler and cylinder and run a heated towel rail from this.

Even if you have central heating, this method has several sound advantages over plumbing in a radiator in the normal way. The rail stays hot even when the heating is switched off in the summer. And in most houses, the pipes between the boiler and the hot water cylinder pass close to the bathroom, so installation is easy and relatively inexpensive to carry out.

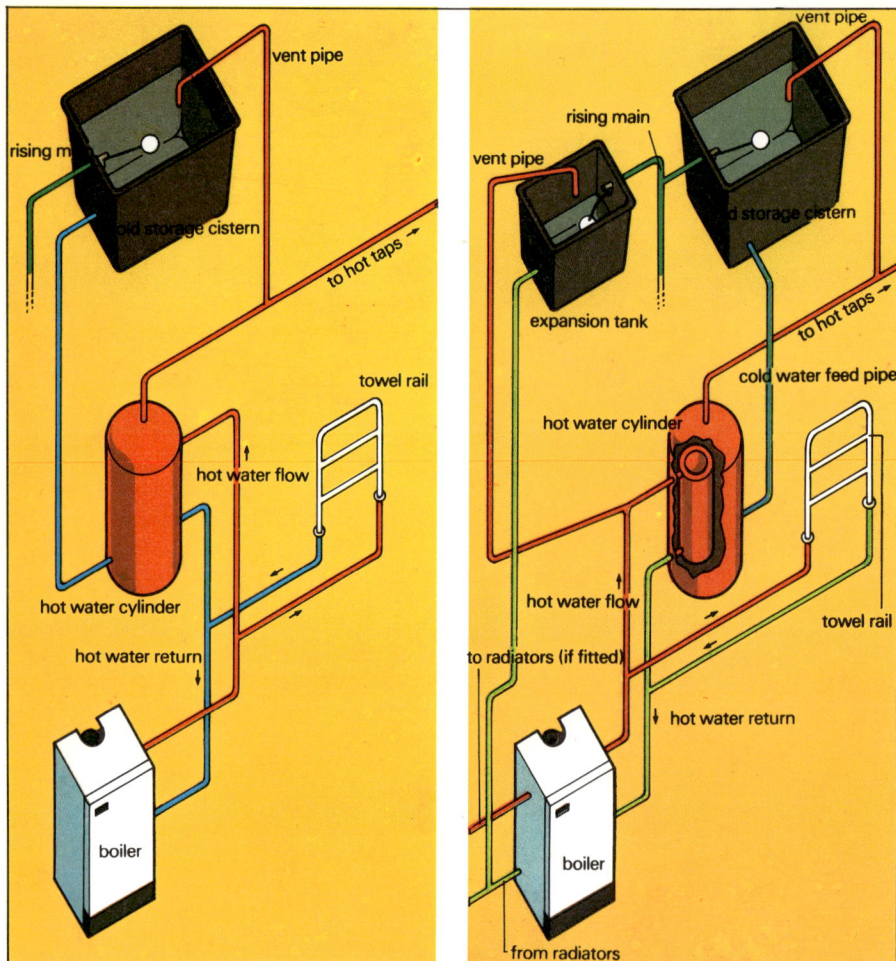

A normal hot water heating system consists of a boiler on the ground floor connected by flow and return pipes to the hot water cylinder on the floor above. The water in the cylinder constantly circulates through the boiler, being heated as it passes through—this is known as gravity flow and is entirely automatic (left).

If a radiator is connected across the flow and return pipes, then the water will circulate through this, too—warming the radiator and so the bathroom.

The system show (left) is the direct heating type—the water that circulates through the boiler is the same as that which comes out of the hot taps. You cannot use normal steel panel radiators with this sytem: they would quickly corrode. You can use only heated towel rails which are designed for connection to a direct system—these are made of copper and therefore you will find that corrosion isn't a problem.

The system shown (right) is the indirect heating type—the boiler water circulates through a coil in the cylinder, heating the water that comes out of the taps indirectly —wet central heating systems use this arrangement, but you may also find it on a hot water only system.

Once filled, the boiler water is more or less unchanging, so corrosion isn't a problem. You can use any form of radiator, radiator/rail or heated towel rail, connecting it across the flow and return pipes as shown.

You may have a fully pumped indirect system—there is a pump on the flow or return pipes. You must fit the towel rail connections before the pump in either case—on the boiler side of the flow pipe, on the cylinder side of the return pipe.

Planning ahead

Your first job is to find out what plumbing system you have, so you can decide which types of bathroom heater you can install. The connections at the water cylinder won't help you much—they look the same for direct and indirect systems. You must look elsewhere for other clues and signs.

If you have wet central heating you can assume you have an indirect system, unless the heating system is very old. You can also assume the system is indirect if it has an **expansion tank**—a small cistern (in addition to the cold water cistern) usually in the loft, connected to the hot water flow and return pipes as shown above.

This leaves systems without central heating or an expansion tank. These will usually be direct, but could use a form of indirect cylinder called a **single feed**, of which the Primatic is the best-known brand in the UK. If it isn't labelled as such, the only way you can be sure is to unscrew the immersion heater after draining down the system (see Draining down, page 50) and peer inside with a torch—any sign of tanks or pipes within the cylinder means it's the indirect type.

Now trace the thick flow and return pipes between the boiler and hot water cylinder. Note the points where you could easily cut into them to form branches: somewhere accessible, and preferably just below the level of your proposed heater. The easiest place is likely to be near the hot water cylinder where the thick flow and return pipes enter and exit from the cylinder (see left). These will generally be in the airing cupboard next to the bathroom itself.

Aim to put the towel rail (or radiator) close to the point you intend to have the branch connections—at the least, ensure the run has as few bends, twists and horizontal sections in it as possible.

Some 'fully pumped' systems have a pump on the flow or the return pipes of the boiler to help the circulation of hot water around the entire system. It's wise to connect the pipes **before** the pump—on the boiler side of the flow pipe, on the cylinder side of the return.

There may also be a fixture called a 'diverter valve' that decides whether to send the hot water to the radiators or to the hot water cylinder. It's important to connect the towel rail pipework **before** the diverter so that it receives hot water whenever the boiler is working. If you're in any doubt, always seek professional advice.

Check the size of the boiler pipes. They are likely to be at least 28mm diameter and could be 35mm; modern, fully-pumped central heating systems may use 22mm (it's easiest to measure the circumference: 28mm pipe is 89mm in circumference; 35mm is 110mm; 22mm is 70mm).

What you need

Apart from the towel rail, the major item you need is piping. The usual choice for this is 15mm pipe, but if you use a radiator

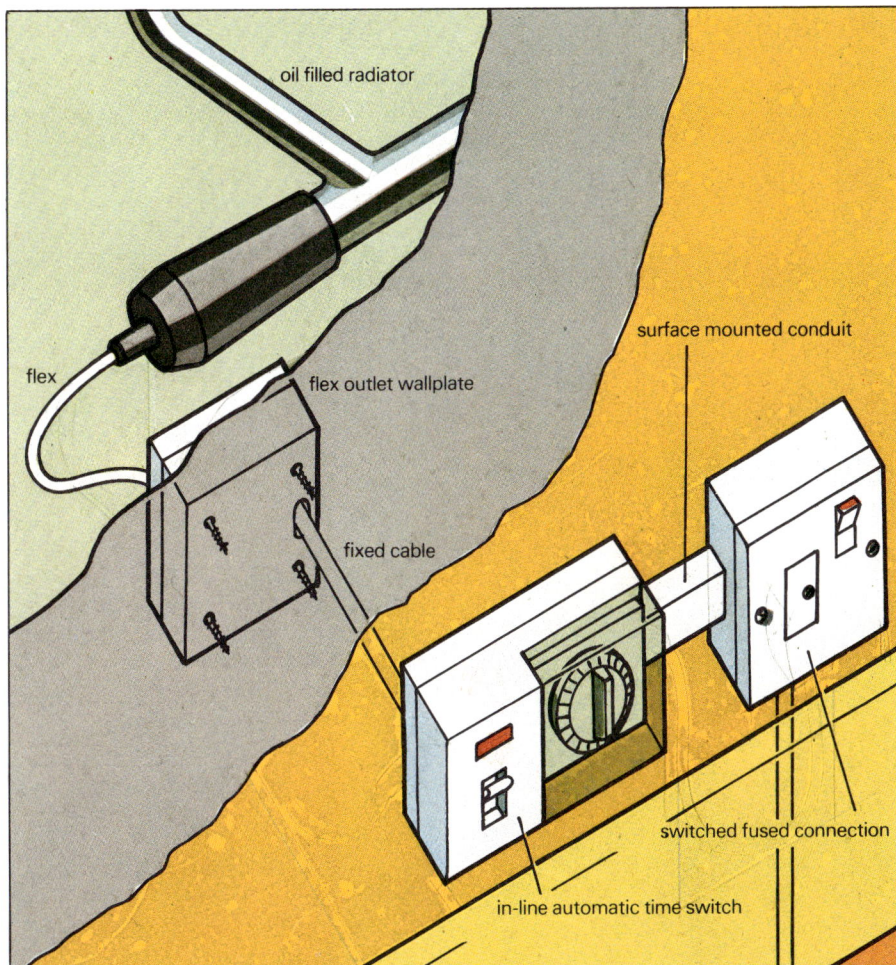

oil filled radiator

flex

flex outlet wallplate

fixed cable

surface mounted conduit

switched fused connection

in-line automatic time switch

it joins directly to the towel rail or runs along skirtings. But if you don't use it, you could paint the copper pipe to match the colour of your skirting boards instead.

★ WATCH POINT ★

Check the towel rail in the position you want it before screwing it home. Make sure there are no obstructions, such as joists, which might make fitting the pipework more difficult than it need be. Nails will indicate the position of the joints below.

Fitting the heater

It's easiest to start at the end, by first fixing the towel rail firmly into place.

On a timber-framed wall, make sure any wall mounting points will be into studs, not hollow cavities.

Towel rails are mounted in various ways, though it's usually fairly straightforward. Some are screwed to the wall and floor using mounting plates. Others are mounted like ordinary radiators—you fix brackets to the wall, then hook the fitting itself over these. Measure off both the brackets and the hooks on the towel rail carefully, so you can work out exactly where to fit the brackets: there is only a little room for error here.

On timber-framed walls, you should try to screw into the studs, because towel rails are heavy fixtures. But with ones that sit on the floor you can use cavity wall fittings.

Electric oil-filled radiators are connected directly to the main supply. In the UK, you must wire the radiator to a proper fused connection unit. The unit can be connected into the power circuit in the same way, and following the same rules, as for an ordinary power outlet.

You can't have a switch inside the bathroom where it could be reached by anybody in the bath, so it's best to switch from the outside. Fit the switched connection unit outside and a flex outlet wallplate inside. An in-line automatic time-switch provides automatic control.

Alternatively, use an unswitched unit and a pull-cord switch

rather than a towel rail, and if it is a long way from the boiler pipework (more than three metres, say) a better choice is 22mm pipe. To take advantage of this, you would need a radiator with larger than normal connections: ask your supplier for guidance on this. Copper pipe is the most widely used, but in the UK plastic pipe can sometimes be used as well.

You will need tee connectors to form the branch with the boiler pipes. You cannot

use the easy-fix types of connector because they are unlikely to be available in large enough sizes to allow the circulation water to flow freely enough under gravity. Again, because boiler pipes are so big, you probably will not be able to use the modern plastic and push-fit types of connector. In most cases, ordinary **compression tees** are what you will need—with 28mm ends (or whatever your boiler pipes are) and 15mm in the branch. If you have, and can use, a blow lamp you could use soldered **capillary tees** instead. After this connection, you can change to plastic pipe and fittings if you find them more convenient.

At the towel rail end, use radiator valves: make sure these suit the connections on the rail and will mate with 15mm pipe. The easiest type to fit can be screwed into the radiator using an ordinary spanner; other types need a large Allen key for this job.

In between, you should need hardly any fittings—you will have to decide where, and which ones, for yourself. Pipes that run along walls or parallel to joists will need supporting frequently with clips.

Stainless steel 15mm pipe might look neater at the towel rail end—at least where

1 *Fit the wall brackets to the towel rail then mark fixing holes at wall or floor level for screws and wallplugs*

2 *Adjust the position to avoid floor joists, then drill, plug and fix at floor and wall. Use a spirit level to make sure the towel rail is upright*

Routeing the pipework

Careful routeing of the pipework between the boiler pipes and the radiator is vital if you want the job to look good, and the rail to work satisfactorily.

The basic aim you should follow in running the pipe is to ensure that it rises gradually, but consistently, from its connections with the boiler pipes to its connections at the towel rail. Any dips in this circuit could stop the gravity flow circulation of the heating water round this circuit, and could also cause airlocks at the top of the rail.

If the only place you can make the boiler pipe connections is **above** the level of the towel rail, you could still have a satisfactory system—but it would be sensible for you to

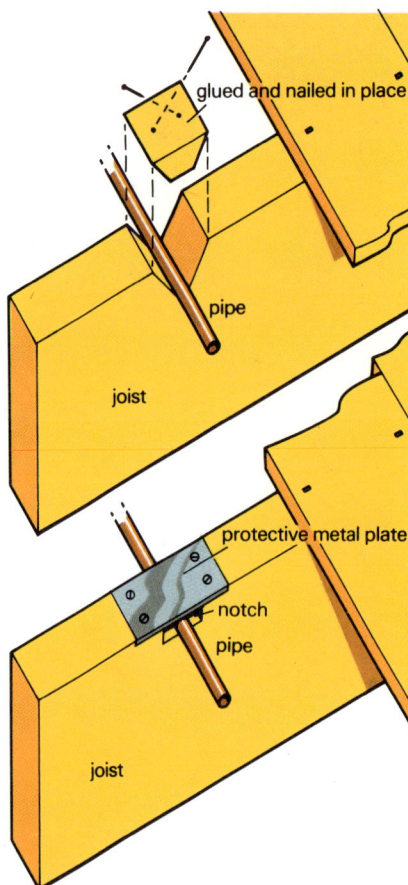

Crossing joists means cutting notches—you may not have room to drill holes

get professional advice first.

The easiest arrangement is to run the pipework along the wall at skirting board level. The amount of rise you need is not very great: a few millimetres per metre run is usually sufficient—so the pipe runs, even though exposed, will still look fairly neat. It is best at this stage not to fix the sections of

pipe permanently together—fit only the main lengths of pipe, clipping them to the wall with plastic pipe clips. Leave yourself plenty of leeway to work round the boiler pipe and rail connections. If you have to run the pipes under a suspended floor, then you should still be able to arrange for a rise so long as they run parallel to the joists. Again, clip the main lengths of pipe to the sides of the joists—or at least support them underneath with carefully driven nails.

Running pipes **across** the joists is a little more tricky. You must make notches in the tops of the joists to take the pipes, and it is often difficult to make these so that the pipe lies snugly with the correct rise in it. One answer is to cut the notches a fraction deeper than needed, and pack out to the correct depth with pieces of felt or old carpet. Packing round the pipes like this is a good idea in any case—it helps reduce the creaking noises when the pipes heat up or cool down. However, take great care not to cut the notches deeper than absolutely necessary, for it weakens the joists.

★ WATCH POINT ★

Making the preliminary connections to the towel rail—called **dressing**—is often easier if you do it before fitting the rail in place. It also makes it easier to see exactly where the pipes will run before you cut the floor.

Making connections

The bulk of the work of installing a towel rail is in making the connections at either end of the pipe run.

Make the connections at the towel rail end first—then you will only need to interrupt the existing plumbing for a short time.

Radiator valves usually come in two parts. One part is a straight connector which screws into the towel rail or radiator bottom connections. Wrap a good measure of PTFE tape round the screw thread, and then screw tightly home. At the top of an ordinary radiator, there are other connections to be made up. One is a blanking plug, simply to fill in an unwanted hole; the other is an air vent plug which has a bleed nipple in it which you can open to allow trapped air to escape. These are similarly screwed in place after wrapping PTFE tape round the thread. You will almost certainly need a large Allen key for these.

In some cases, you will be supplied with

3 *Underfloor routeing is best—try to run the pipes parallel to the joists. Use pipe clips to secure the pipes along their length, preferably at 450mm intervals*

4 *If it's inconvenient to clip the pipes to the sides of the joists, use battens as shown to support the pipes between them. Allow a small gap for expansion*

5 Fit the valves to the inlet and outlet holes of the towel rail—use PTFE tape on the connector only so that you can adjust the position of the valve to meet the pipework

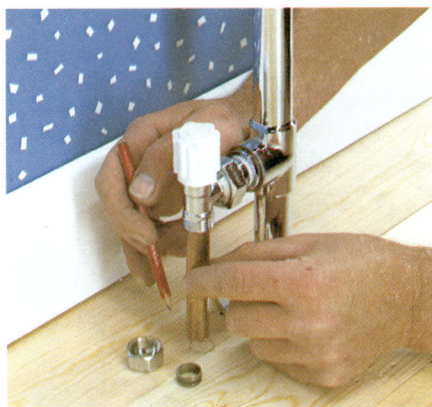

6 Mark the floor where you need to drill holes for the pipes—use an offcut of pipe to mark the hole position and make sure that it is immediately below the valve opening

7 Remove the valve itself to drill the hole. Use a 16mm flat bit and power drill. Hold it upright to avoid damaging the valve connector when the drill breaks through the surface

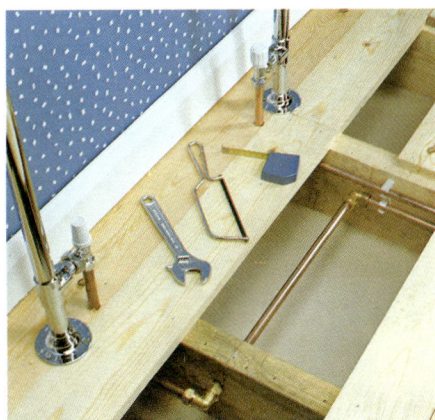

8 Start the connections at the towel rail end. Assemble the first elbow joint and slide the pipes into place. Connect one end to the valve and support the pipes with clips

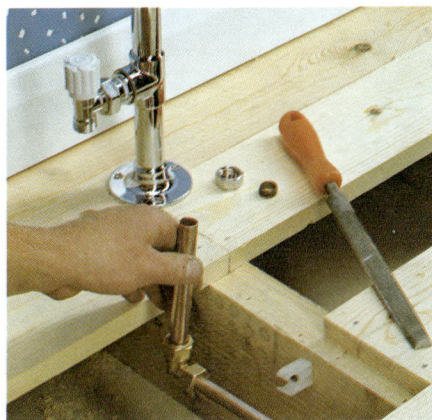

9 Connect the other end of the towel rail in the same way—direct the pipework along the same route as the first. You'll need to cross a joist to do so (see Routeing pipes on page opposite)

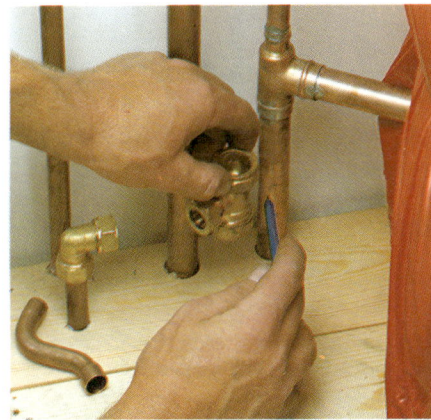

10 At the cylinder end, drain the pipes and mark them for the connectors on the flow and return pipes. You may need to use special 'crossovers' to avoid obstructing pipes

11 Cut through the pipes with a hacksaw and deburr the ends, inside and out. Fit the connectors—you may need to loosen the pipes in order to slide them into place

12 Connections at the boiler/cylinder circuit should look like this—the flow pipe connection should be higher than the return pipe connection to aid the circulation

two valves—one control valve for the flow into the towel rail and one for the return out of the towel rail. Make sure you fit them the right way round—if necessary, label the valves and pipework to avoid confusion.

The radiator valves themselves fit on the end of the straight fitting—the connection here, which is part of the fitting, is a type of compression joint: you can do it up loosely and still move the valve around to line it up with the pipework. When making the joint, smear the metal mating surfaces with a coat of jointing compound or PTFE tape. The other end of the valve has a normal compression joint to connect it to the pipe. Connect this one loosely by tightening by hand—without jointing tape or compound —until you're absolutely ready to make the rest of the connections to the flow and return pipes on the radiator.

At this stage, you should be ready to connect the whole new system to your boiler pipes. It's also worthwhile running through the job to anticipate problem areas such as changes in direction and boiler pipe connections.

First, drain down the relevant part of the system (see this page). Then begin connecting the pipes between the towel rail and the boiler pipes (see page 48). Work from the towel rail back towards the boiler/cylinder circuit, using any type of joint you prefer—there's really not much difference between them on a job like this. Compression joints are generally the easiest to make—especially for underfloor routeing where there's the risk of starting a fire.

With your tee-fitting in hand, measure the amount that you will have to cut out of the boiler pipework—this is the length of the body of the fitting less **twice** the amount that the pipe will disappear into the connector when it is inserted.

Mark this distance on the boiler pipes at the exact point that you want the towel rail pipes to branch from, and cut out the section of pipe carefully with a hacksaw.

Your next problem is going to be inserting the long body of the fitting into the short gap you have cut out. There may be some 'slack' already in the pipes, or you may have to remove some pipe clips to give you a bit of leeway. In extreme cases, you will need to dismantle a whole section of pipe by unscrewing at the nearby compression joint.

Before tightening your new tee, connect the branch run to it, and make sure all the pipework aligns neatly and without strain. If the pipes don't lie easily and loosely in the fitting, the chances are that the joint will not be watertight. Faults like this are easier to remedy now than at a later stage, even if it means you have to remake a whole section of pipework.

Double-check your joints; refill the system; and test with cold water before relighting the boiler. Then heat the water, and check again—keep your eye on the joints for a couple of days after installation: leaks may take a little time to show up.

> ★ WATCH POINT ★
>
> Before starting work on the existing system, double-check that the rest of your installation is finished and sound. Then check that you have all the tools and materials that you might want for the next stage.

Draining down

The procedure for draining the boiler pipes varies a little depending on whether you have a fully indirect system or not.

With a fully indirect system you should first switch off the boiler and let it go cold.

Cut off the water to the expansion tank by closing the valve, or tie up the ball valve using a stick and a piece of string.

Locate a drain tap on the lower boiler pipe, usually close to the boiler. Attach a hose to the drain tap and lead it outside. Open the tap and let the system drain for a few minutes.

The hot water cylinder will remain full, and if you have an immersion heater you can still use it to provide hot water for baths or for washing up.

With any other sort of system you follow the same procedure, except that you cut off the water by closing a valve on the pipe between the cold water storage cistern and the hot water cylinder. If you haven't got a valve on this pipe, you must cut off the flow into the storage cistern, or tie up its ball valve. You will be emptying the hot water cylinder, so you will have no hot water until you refill. And, if you have to cut off the flow to the storage cistern, you will have no cold water (except in the kitchen).

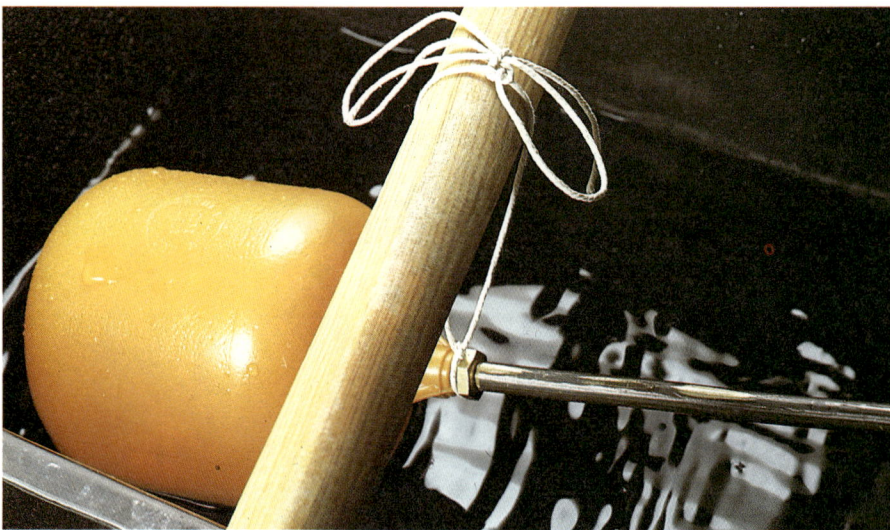

13 Tie up the ball valve at the expansion tank (or the cold water tank in a direct system). This will prevent the tank from refilling when drained

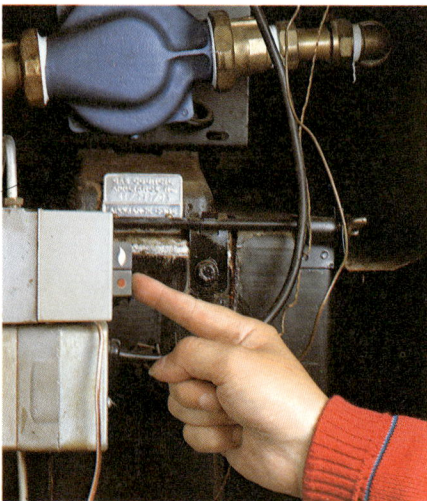

14 Switch off the boiler and allow it to cool. With an indirect system, you can still use the hot water but with a direct system you should also switch off immersion heaters

15 Drain the boiler. Attach a hose to the lower drain plug and unscrew the plug. Allow the system to drain down—it may take some time to clear all the pipes, so be patient

CHAPTER 3

UTILITY AREAS

Plumbing in a washing machine or installing an outside tap can greatly improve facilities in your home. While both sound like simple jobs, there are plenty of pitfalls for the unwary. So tackle the work systematically and never try to cut corners.

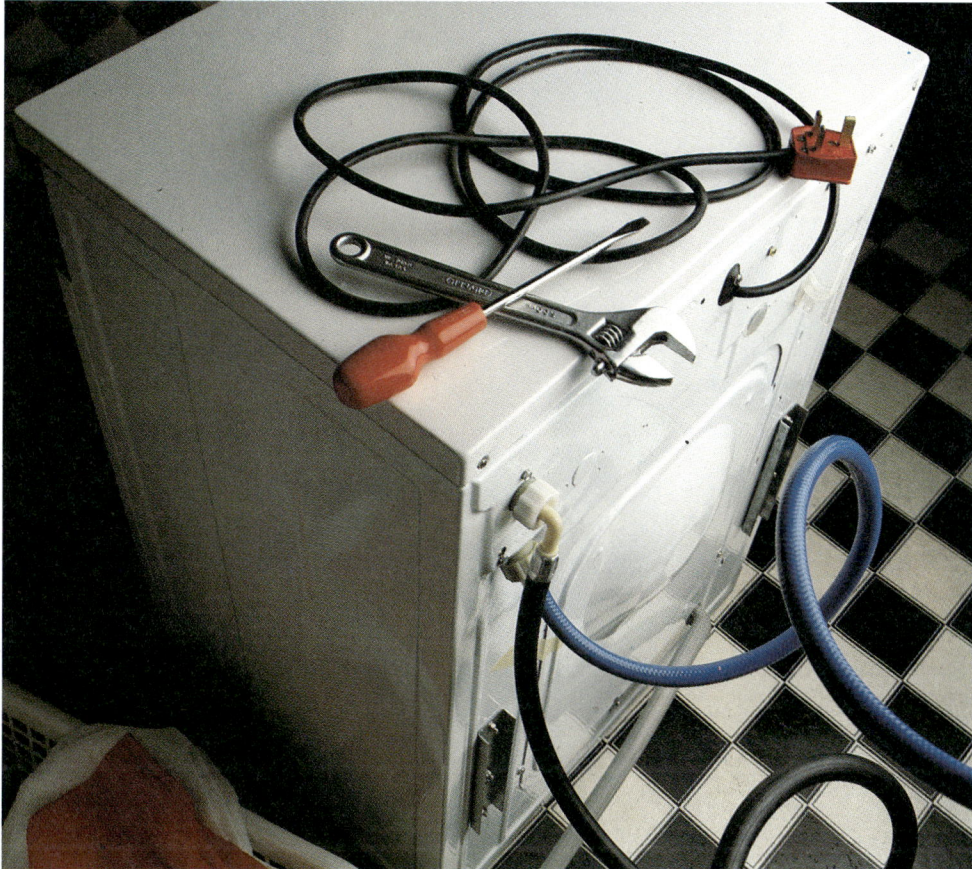

FITTING AN OUTSIDE TAP

Having an outdoor tap puts an end to traipsing through the house every time you want to fill a bucket or attach a hose. And as well as being an invaluable gardening aid, it's a boon if you're doing regular building work or regular cleaning, like washing a car. If you invest in an automatic hose reel as well, you can fit this next to the tap, ready-connected, for greater convenience.

As plumbing jobs go, installing an outdoor tap is about as simple as it could be: you've only got one supply pipe to worry about, plus the fairly easily solved problem of piercing a hole in an outside wall. Using conventional plumbing materials and techniques the job's straightforward enough. But there are complete kits on the market, similar to those for plumbing to a washing machine, which are designed specifically with the do-it-yourself enthusiast in mind.

At their simplest, such kits consist of the tap, wall fitting, stop valve and supply connector. But the latest include almost everything that you're likely to need, including easily workable plastic supply pipe, push-fit connectors for jointing, and an automatic supply connector for breaking into an existing pipe without having to turn off the water.

Planning and preparation

When choosing a site for an outdoor tap, bear in mind the following points:
• The tap should be sited where it's going to be of most use. For example, if you need to use a hose front and back, don't position the tap so far round one side of the house that it's impossible to get the hose to the other.
• Try to site the tap over a gully, pathway or at the very least hard ground: there's bound to be some spillage and grass or soil could quickly turn into a mud patch.
• For maximum water pressure, the tap should be connected to the rising main. This means finding out where the main runs, and then reconciling the most convenient connection point with the ideal site for the tap to give you the shortest possible pipe run.

The rising main is usually easy to find because a branch from it will run directly to the kitchen cold tap—this rule applies irrespective of whether your plumbing is direct (when all fixtures are fed direct from the main) or indirect (in which case you'll have a cold storage tank).

Starting under the sink, trace the route of the rising main and make a note of possible connection points along the way. When you find what looks like the best spot, check that from there to where you will break through the wall gives you enough room to fit an unobstructed pipe run with stop valve and sufficient space to drill through the wall itself. This isn't as obvious as it sounds: there's nothing more frustrating than starting the job only to find that you can't complete it without tearing out half your kitchen fixtures.

When you've settled on a location for the tap and chosen a suitable connection point on the rising main, you're almost ready to start. But before you buy any materials, get in touch with your local building control office or water authority. In some areas, work on the rising main must be left to an authorized plumber; in others, there may be bylaws specifying what materials you can use—for example, some authorities still insist that you use metal pipe and soldered joints. You'll almost certainly have to pay an extra charge on your water rates for the extra tap.

Tools and materials

If you buy a kit you get all you need except the extension pipe (see opposite). If you decide to opt for individual components rather than a kit, purpose-made garden taps are available from builder's merchants and garden centres. They're normally made of brass for maximum weather resistance, have crutch-type handles for ease of operation, and should come complete with wall brackets and fixing screws.

As well as the tap itself, you must fit a stop valve on the supply pipe inside the house so that the tap can be isolated and drained in the event of a frost. Traditional stop valves are all brass with compression or capillary connections at each side. Plastic pipe manufacturers include push-fit jointed valves in their ranges.

Subject to local water authority restric-

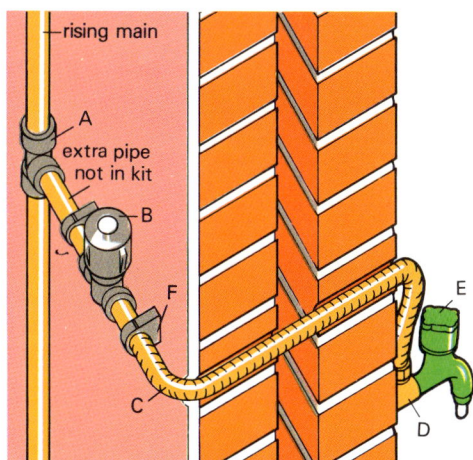

An outside tap kit contains: push-fit tee (A); stop valve (B); pliable copper pipe (C); wall elbow (D); tap (E); pipe clips (F)

tions, the connection at the rising main can be made with a capillary, compression or plastic tee fitting, or with an automatic connector; of the latter, choose the water authority approved type, in which the hole is cut by a small firing pin that traps the resultant metal swarf (which can cause damage) permanently in the fitting.

Note that if you connect to an old type rising main, you'll need a 22mm tee reducing to 15mm for the new branch; if you connect to the kitchen cold tap branch, buy a standard 15mm tee.

You need enough 15mm pipe to make the connection, plus elbow and slow bend fittings and copper bendable connectors in your chosen jointing system to take care of bends. You also need a wallplate elbow tap connector to connect the new tap to the wall and to its supply.

Pliable copper pipe can be run directly through a 15mm hole in the wall—seal around it with mastic. Alternatively, you can arrange a protective conduit through the wall: 22mm plastic pipe is best for this —buy about 300mm, plus non-setting mastic and repair mortar or exterior grade filler to make good the hole.

Tools for the plumbing work are straightforward: a hacksaw and tape measure for plastic pipe, plus a pair of wrenches, grips or adjustable spanners if you're making compression joints. For capillary joints you'll need a blowlamp, resin-cored solder, flux, wire wool and a heat-proof tile.

Running pipe through the wall

Cutting a hole in the wall to take the supply branch pipe is the first, and most daunting,

part of the job. But if you measure up properly you can tackle any sort of wall with confidence.

You can cut the hole for the pipe using an electric hammer drill fitted with a heavy duty masonry bit and extension (the bit will have to be 15mm or 22mm in diameter). Or you can hack through the wall with a club hammer and long cold chisel.

Unless you're using a very long drill, you'll most likely have to work from both sides of the wall. It's therefore important that the holes are aligned exactly.

Start by marking on the outside where the tap is to go, then select a reference point that will be visible from the inside as well—a

window or door are the most obvious.

Measure from the reference point to the tap location in straight horizontal and vertical lines, using a try square or spirit level to guide you. As you go, make a note of the measurements. Then move back inside the house and repeat the measuring procedure exactly, working from the same reference point. Where you end up marks the point at which you drill or cut through the wall.

If you are using a drill, set it on the lowest speed setting and start drilling from the inside to keep damage to the plaster to a minimum. Make sure you drill at right angles to the wall.

1 *Drill a hole through the exterior wall to take the pipe using a hammer drill with a large diameter masonry bit. A number of holes may be needed*

2 *Widen the mouth of the hole so that you can bend the pliable copper pipe to meet the water supply without too sharp an angle*

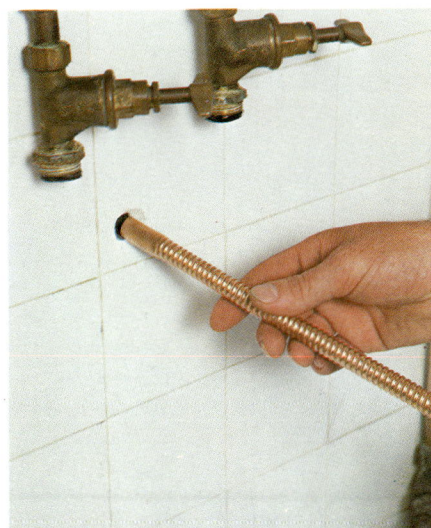

3 *Insert the pliable copper pipe through the hole in the wall. Go outside and work out how much you need*

4 *Hand-bend the pliable copper pipe in the direction of the rising main, being careful not to form too tight an angle*

On a cavity wall, you'll feel the drill suddenly free itself as you break through the inner leaf. If you haven't a long extension piece for the drill, stop at this point and drill through from outside. On a solid wall, stop when you've drilled to a depth of approximately 110mm.

Using a hammer and cold chisel, your priorities are to keep the hole straight and as near 22mm as you can. As when drilling, start with a hammer and chisel from the inside. You'll cut down damage to the plaster if you start off by drilling a series of holes through it using an ordinary masonry bit. As you get into the wall proper, stop every so often and clear out the debris with a piece of wire.

However you break through the wall, don't forget to wear goggles so that you are protected against flying chips of masonry.

When the hole is finished, take your piece of pipe (or conduit) and try it for fit. Drill or chisel out more masonry where needed. With conduit, re-insert the pipe in the hole and cut it with a hacksaw to the exact thickness of the wall plus plaster.

Finally, make good any damage around the hole inside and out with repair mortar or filler. This should leave you with a neat hole or conduit through which to run the new supply pipe to the outside tap.

Connecting to the rising main

There are four ways of doing this job—with a soldered or compression tee, with a push-fit tee, or with an automatic connector.

Before you do anything else, double check that you've got the right pipe and mark on the connection point in felt-tip pen so that you don't get confused later.

Conventional tee branch

Drain down the rising main by turning off the main stop valve and opening the kitchen cold tap. Place a bowl under the connection point and then cut through the pipe using a junior hacksaw. Take care that you keep the cut absolutely square.

Measure and mark along from the cut the

width of the fitting; don't forget to allow for the amount that will slot inside it at both ends. At this point, make a second square cut through the pipe. Afterwards, file off the burrs on the cut pipe ends—inside and out—and in the case of a compression tee shape the edges to a slight bevel.

5 *Hold the push-fit tee against the rising main and mark off on the pipe how much you'll need to remove; this is denoted by ridges on the tee*

7 *Make the second cut through the rising main and remove the offcut. Prepare the cut pipe ends for fitting into the tee*

Before you fit the tee, you may need to remove one or more pipe brackets so that you can ease it into the gap.
Compression tee: Slip the capnuts onto the pipe ends, followed by the olives, which should go about 12mm down the pipe.

Insert the pipe ends into the fitting, screw on the capnuts and tighten about one and a half turns above hand-tight using a pair of adjustable spanners or wrenches.
Soldered tee: Clean the pipe ends with wire wool until they're shiny, then smear them with flux. Slip the pipe ends into the

6 *Drain down the rising main by closing its stop valve and opening the cold taps; cut through the pipe using a junior hacksaw*

8 *Loosen the pipe clips retaining the rising main and slot on one side of the push-fit tee. Ease the other cut pipe end into the socket*

tee. Place a tile or heat-proof board behind the pipe run to protect the surrounding area from the blowlamp flame.

Run a blowlamp over the first joint until the flux starts to bubble, then ease off slightly. When a small ring of solder appears right around the edge of the fitting, cut the flame and leave the joint to cool.

Plastic push-fit tee: Lubricate the pipe ends and the inside of the tee, then slot it on the pipe run.

Automatic connector

This method is simplicity itself—you don't even have to turn off the water.

Clamp the connector over the pipe at the connection point and tighten the four clamp screws using the Allen key provided. Check that the outlet hole is pointing in the direction of the run.

This is literally all you do at this stage. The charge which cuts through the main is activated when you've assembled the run.

Installing the pipe run and tap

This final stage of the job is perfectly straightforward, providing you take care to make each joint properly—in this way you will avoid leaks.

When installing the pipe run and tap, take your time. Professional plumbers may work quickly but they generally have years of experience behind them.

Conventional copper plumbing: Measure, mark and cut sections of pipe to allow for the inclusion of a stop valve at a convenient point. If you need to start with a bendable connector at the tee, fit this and bend it to shape, then measure on from here as far as the stop valve. Before you assemble the run, pin pipe brackets to the wall at least every 1m, plus one each side of the valve itself. Then make compression or pre-soldered joints, according to preference.

9 *Push the stop valve on to the prepared end of the pliable copper pipe. Mark the wall to take a plastic pipe retaining clip*

10 *Swing the pliable pipe and stop valve out of the way and screw a plastic pipe clip to the wall. Clip the pipe in place*

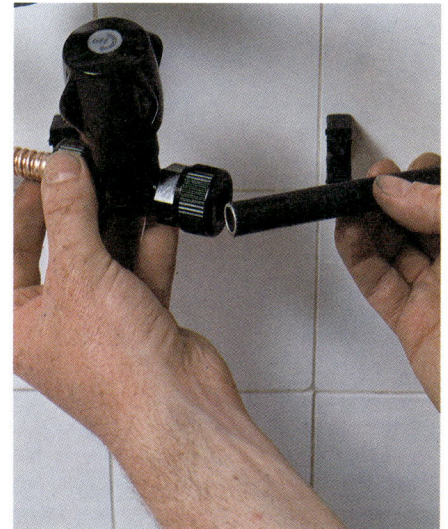

11 *Measure the distance between the stop valve and tee fitting, allowing for the small amount to be inserted into each of the fittings*

From the stop valve, measure and cut enough pipe to take you to the elbow that turns the run through the hole in the wall. Then, before you fit the elbow, cut another length of pipe to run through the wall so that it protrudes about 25mm on the outside. Fit this to the elbow before fitting the elbow to the rest of the run. Check that the pipe run is held rigidly in its brackets and that all the joints are properly made, then move outside.

Solder or compression-joint a tap connector to the bare pipe end, then check the tap fitting instructions. If the tap has a bracket, mark the fixing holes, then drill and plug

12 *Cut a length of flexible plastic pipe to size and push a special metal insert into each end. Insert the pipe tightly between the fittings*

the wall. Fit the tap to its connector simply by tightening the capnut with an adjustable wrench or spanner. Where necessary, secure the tap bracket to the wall with the screws provided.

Plastic plumbing (push-fit joints): Start off the run by fitting a straight tap connector or bendable tap connector to the automatic connector outlet or tee on the rising main.

From here you simply assemble a run of push-fit-jointed plastic pipe including a stop valve and a 90° elbow fitting to take it through the wall. Cut the sections of pipe running through the wall to protrude 25mm on the outside and fit it to the elbow

13 *Outside, slot the compression wall elbow onto the copper pipe. Fit capnut, olive then elbow body and then tighten the capnut*

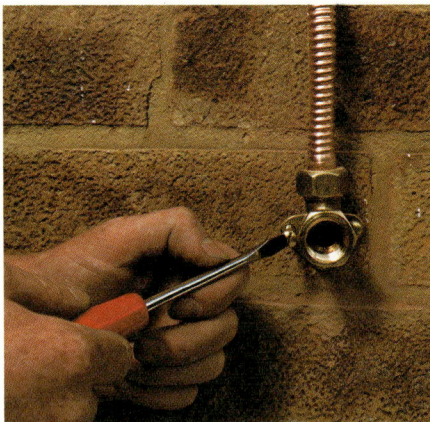

14 *Mark, drill and plug the wall for the elbow fixing screws then fasten it to the wall using the screws which are provided in the kit*

15 *Finally, screw on the bib tap to the end of the wall elbow. You may need to wind on PTFE tape so the tap will be vertical*

before joining the elbow to the rest of the run. Secure the pipe run with clips every 1m, plus two each side of the stop valve (see Plastic pipe and push-fit joints). Some kits include pliable copper pipe instead of plastic.

Now move outside and fit a push-fit straight tap connector to the free end of the pipe. Drill bracket holes for the tap itself (if needed) then screw the tap to the connector.

The final job with plastic pipe is to activate the automatic connector on the rising main. Check that the new tap is turned off, then remove the plastic firing pin cap on the fitting and strike the firing pin sharply with a hammer: you'll hear a loud crack when the charge goes off to pierce through the pipe.

Check carefully for leaks around the automatic connection. Retighten the nuts around the connector if necessary. Also check the tightness of the other connections.

Plastic pipe and push-fit joints

Assembling a run of plastic pipe with push-fit joints is very easy if you follow a few simple rules:
• Cut the pipe squarely using a hacksaw and a piece of card as a template. Afterwards, remember to smooth off any burrs on the cut edges using the emery cloth provided with the pipe or by using the de-burring tool fitted to the end of a pipe cutter.
• Smear the cut end to be joined with silicone lubricant (also provided) before pushing it home into the joint with a slight twisting motion. Make sure the pipe goes right into its seating.
• If space is too limited to push the pipe home, unscrew the joint fitting. Slip the cap over the pipe end, followed by the grab ring and O-ring seal. Then screw the cap down tightly to the other half of the fitting.
• Don't disturb the joint once you've made it. If you do have to take it apart, unscrew the fitting and fit a new grab ring and seal prior to reassembly.
• When using copper bendable connectors, avoid bending them more than once and keep the bend to as large a radius as possible. If you have to cut the connector, do so on a plain section only and leave 25mm for connection to the next push-fit joint along the run.

Final moves

On a conventional installation, start by re-instating the water supply. Then, in all cases, operate the new tap and check the entire run for leaks. Test the stop valve, too, which you should close if a frost is expected.

Finish off by filling the small gap between the new supply pipe and its plastic conduit —inside and out—with non-setting mastic.

Push-fit joints can be used with either plastic or copper supply pipes

Deburr the ends of copper pipe with emery cloth or wire wool

Lubricate the pipe end with a little silicone lubricant to help insertion

Lubricate the inside of the fitting and insert the pipe firmly

Flexible plastic pipe needs a metal supporting insert for fixing

WASHING MACHINE PLUMB-IN

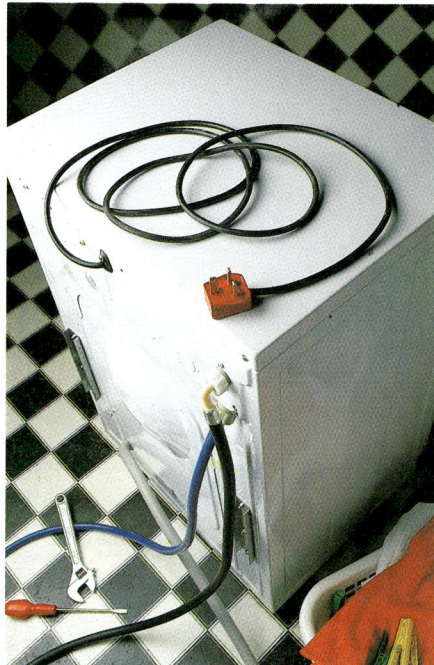

The complicated part of this job is deciding where to put the machine in relation to the available plumbing fittings but this can be made easier by approaching it in a logical order. The actual plumbing work is usually simplicity itself thanks to the wide range of special plumbing fittings for the do-it-yourselfer that have come on to the market in the last few years.

Siting the machine

It isn't only the space you have available which dictates where you put your washing machine. Just as important—more so in many cases—is the ease with which you can connect it to the supply pipes, your drainage system, and a power supply.

Consider these problems one by one, bearing in mind the plumbing methods you feel happiest about (see below) and also any local authority restrictions that are in force in your area.

The supply pipes

All washing machines need a cold water supply, and many take a hot water supply too, so it follows that the ideal site is close to the relevant water pipes—however you decide to break into them. Many people like their machine to go near the sink or in a washroom with exposed pipework: not only

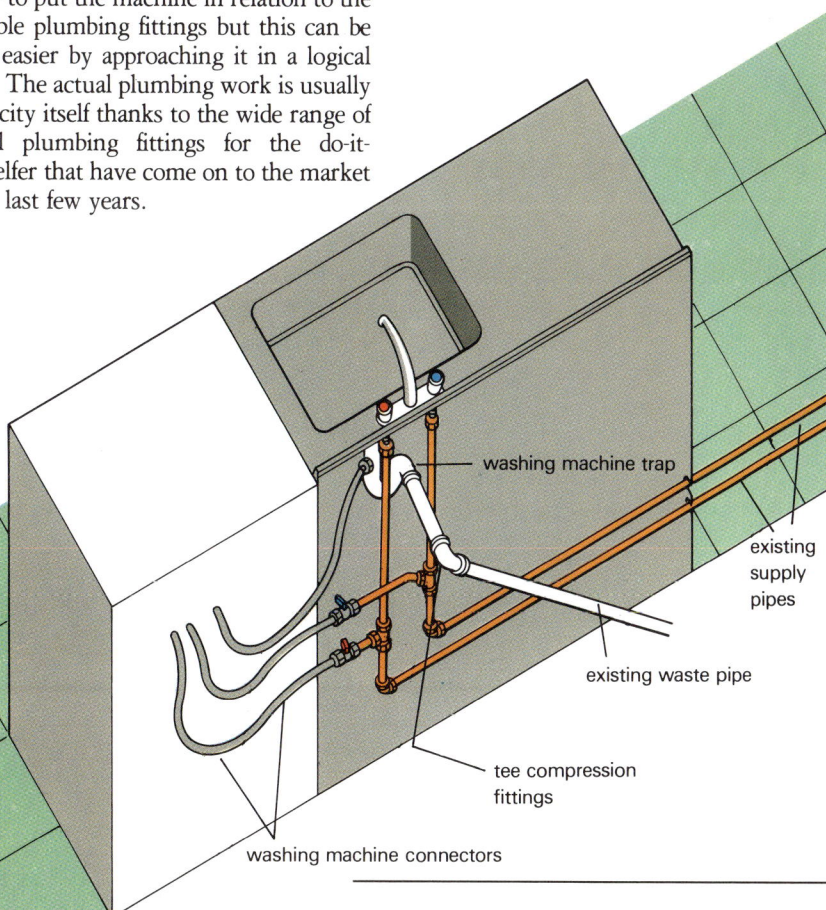

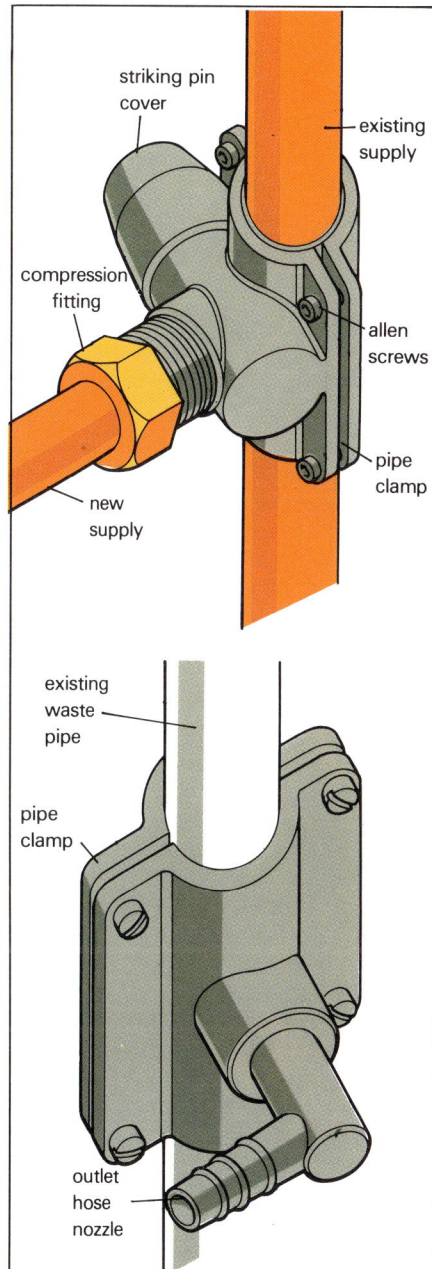

Left: close connection is the simplest to arrange—thanks to the wide range of easy-to-use joints and fittings. The installation shown uses tee branches in the supply fitted with washing machine connectors. The waste is plumbed to a washing machine trap. Alternatives (above) include automatic supply connectors. You can also get supply connectors which include stop valves

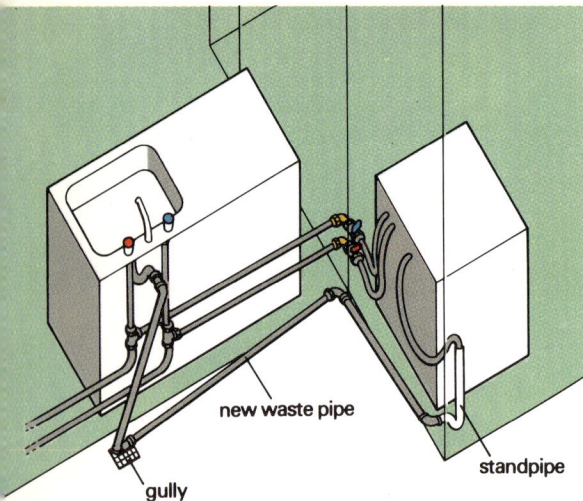

new waste pipe
gully
standpipe

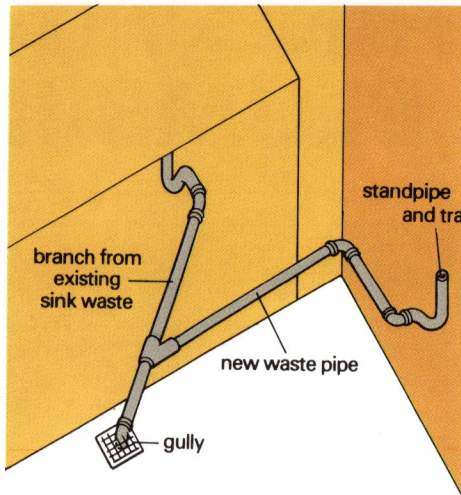

branch from existing sink waste
new waste pipe
gully
standpipe and trap

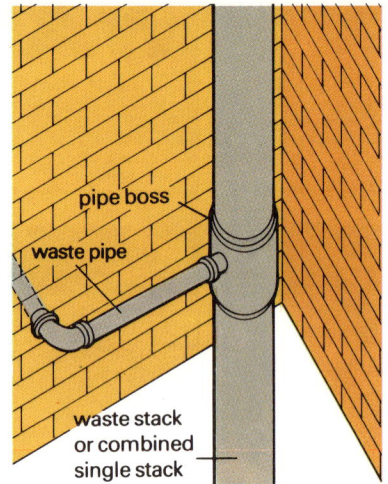

pipe boss
waste pipe
waste stack or combined single stack

Above: remote connection is not quite as simple as the close connection shown on page 57—because of the complication of extending the pipes. The supply connection is identical—except that the pipe run is longer. But if there is no adjacent waste, you need to connect via a standpipe. This can run directly into a gully, or (right) join into an existing pipe using a branch connector. On an upper floor, you can run into a hopper head or directly to the stack (far right)

does this give easy access to the supplies, it also puts the outlet within reach of a drainage point. But don't let the proximity of water pipes influence you too much. It's far better to let the most convenient drainage point dictate the site and then extend the supply pipe routes to fit in with this. What you must ensure is that where you choose to break into the pipes gives you enough room to do the job properly.

Telling hot pipes from cold is easy—unless you have central heating. You must ensure that your hot feed is connected to the pipe supplying a hot tap—not a heating pipe. If there's likely to be any confusion, turn the heating off and run the hot taps. Then mark the pipes at the break-in site so that you don't muddle them up later.

On indirect plumbing systems with a cold storage tank, the cold water pipe you've selected might be the rising main or the branch of it that runs direct to the kitchen sink cold tap. In this case, a word of warning: some local authorities do not permit home owners to interfere with the rising main, in which case you must find a supply fed from the tank. Check this point carefully before you proceed any further with the project.

Water pressure: On direct water systems, this is governed by the rising main pressure;

there should be no problem but your water authority can supply details if required. On indirect plumbing systems the pressure is governed by the head—the distance between the base of the storage tank and the washing machine inlets. Normally the minimum head required is 2.44m (8ft). If yours is less than this, consult a plumber about ways of boosting the pressure.

Drainage

Of all the site considerations, this is invariably the overriding one. The reason is simple—you have to fit in with your existing drainage system, or else face a prohibitive amount of work.

However you connect the drain outlet on your machine to the waste system, it should meet the relevant water authority regulations. These are designed to guard against the possibility of back-siphonage—dirty water being sucked back into the machine, with the risk of contaminating the supply. There are two ways to arrange this.

The easiest is to use a special fitting which taps into the waste pipe from a sink or another appliance. The fitting has a valve in it to ensure that the flow is one way—that is, to rule out the chance of back-siphonage into the machine. A variation on this theme is the washing machine trap, which in-

cludes an extra waste inlet and replaces the existing trap underneath, say, a sink.

The alternative method is to connect the machine outlet via a standpipe and trap (to avoid the siphonage problem) to its own waste pipe. This can then be tied into the waste pipe of another appliance or—with more difficulty—connected entirely separately to the nearest available waste outlet.

The main drawback to the first method is that the washing machine must be near an existing waste pipe for it to be practicable. A second point is that not all washing machine manufacturers endorse it, and not all authorities permit it. However, it is the easiest—so it's worth checking up on before you go any further.

The difficulty with the second method nearly always centres around one of two things: either there is no convenient waste to tap, or the drain outlet is untappable. But if you have to run a separate waste you must also bear in mind that it should run in as straight a line as possible, at a gentle downward gradient, otherwise it will be illegal. For details, see below.

Your choice of suitable outlets depends on the exact layout of your drainage system. In a two pipe system you can connect into a gully or a waste stack. The gully option is usually simple, and so is the stack if the washing machine is at first floor level because you can generally run it into a

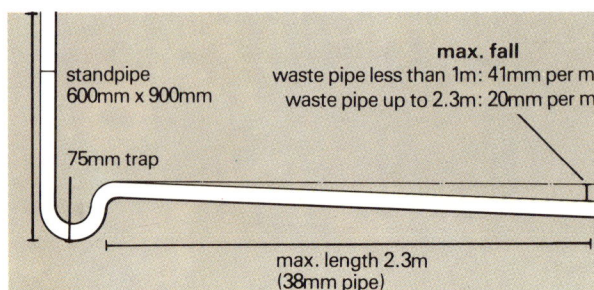

standpipe 600mm x 900mm
75mm trap
max. fall
waste pipe less than 1m: 41mm per m
waste pipe up to 2.3m: 20mm per m
max. length 2.3m (38mm pipe)

A simple standpipe (left) will provide a separate waste outlet. To connect the machine, its hose is pushed loosely into the top—the air gap prevents back-siphonage. Note the dimensions of the waste —particularly its slope or 'fall'

hopper head. Connecting directly to the stack is difficult, and unless there's a spare inlet hole on it, it's not really worth considering for the sake of a washing machine—select an alternative site where drainage poses less of a problem for you to deal with.

Modern single stack systems impose even more restrictions. There are fewer gullies and it's often much harder to break into the stack. Weighed against this, however, waste systems from existing fittings and appliances are generally more straightforward—so you should be able to find an easier option than connection to the drains direct.

As with supply pipes, there are various local authority restrictions relating to drainage—some, for example, are very particular about the way you connect a new waste pipe to a gully. Fortunately, water authorities with tight restrictions are usually very helpful when it comes to offering practical advice. So contact your inspector and tell him what you propose to do—after all, the rules are there for a reason and you may be able to save yourself from making a costly mistake.

Electricity supply

Of all the services, this offers the most flexibility. If you can't run the machine's flex, via a plug, to a nearby outlet, you have the choice of adding a new socket or—better, because it gives the machine its own supply—a fused connection unit (see page 18). Bear in mind though that in the UK if you want to put the machine in the bathroom, the supply point must either be outside the room or else be an approved fused connection unit with pull cord switch.

What you need

When you have decided on a site, familiarize yourself with the different techniques and materials for breaking into the supply pipes and drainage system.

What you eventually buy will depend on the nature of your existing plumbing, the access on and around the site, and the easiest method of doing the job. Often this is a matter of personal taste. Many of the special accessories listed below would be too expensive to use on large-scale plumbing jobs, but on a small project like this they pay for themselves ten times over.

Start by looking at plumbing-in kits, which are now becoming more widely available. Some deal only with the water

For the supply, use a tee joint or automatic connector, plus a washing machine connector

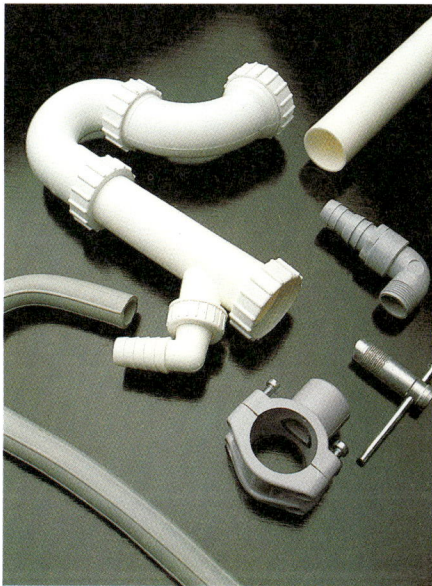

For the waste, there are break-in connectors or washing machine traps which fit onto the sink

supply, some with the drainage and others with both. Getting everything in one go certainly makes life easier but it is only worth it if the parts provided actually suit your site. Check this before you buy; all kits will contain a combination of the fittings described below.

Water supply: On each pipe, you need a tee connector, a stop valve, a washing machine connector to match the inlet hose on your washing machine, and enough pipe or flexible connectors to piece the new run together. These, however, are the basic requirements: they can often be simplified by

using one or more special accessories.

• **Automatic connectors** break into supply pipes without you having to turn off the water supply, eliminating the need for tee branches. Unfortunately, very few of the many types on the market are actually approved by washing machine manufacturers and water authorities (most let the cut-out section of pipe fall into the supply when you first connect them, which can damage the machine). One type that is safe is simply clamped around the pipe at the break-in point. Striking a pin on the fitting then activates a small charge which sends a cutter across the side of the pipe and automatically retains any metal cuttings in the fitting itself.

• **Washing machine connectors** generally combine a connection point with a stop valve in the same fitting. As you must always fit stop valves, this saves a lot of trouble.

As well as accessories, you need to decide what new pipework and jointing system to use. As regards the pipe, you have the choice between copper and plastic. If you opt for plastic, you can buy special adapters for connecting to an existing run of copper. Where necessary, though, check that the plastic pipe is suitable for both hot and cold water. Pipe size for both materials is 15mm.

When it comes to joints, the choice is generally between compression and push-fit. Both are much more expensive than the soldered type, but they are far easier to fit too. And as you don't need many anyway, it's the simplicity that should tip the balance. Do check, however, that any accessories you are buying are compatible with both the pipework and the jointing system you have.

When you cannot connect a washing machine connector/stop valve direct to the supply pipe, it's a good idea to sketch a plan of the new pipe run. Mark in approximate pipe lengths and the points at which you need joints or flexible connectors. List all the items before visiting your plumbing supplier.

Drainage: Again, there's plenty of choice. But in this case what you eventually need will be determined by your break-in method, itself determined by the site.

Automatic connectors for breaking into existing plastic waste pipes are very useful and have more general approval because the section cut out when you make the connection is discharged straight into the drain. You simply clamp the fitting onto the waste pipe—most designs have a sleeve so that they can be used with any size of pipe. Then you make a hole in the pipe using the cutter provided, and screw on the valve nozzle—which connects to the washing machine's outlet hose.

• **Washing machine traps** offer an alternative easy connection method. You simply trim the new trap to match the existing pipework, fit it in place of the original waste trap, then slip the machine outlet hose over the nozzle.

• **Standpipe kits** contain all you need if you want to do the job the other way. But in this case you'll also need elbows, tees and straight sections of new waste pipework to connect to the existing waste, to a gully or to the stack. The diagrams on page 66 provide examples of possible pipe runs. Use these to sketch a plan of your own run, then make a list of the parts required. The waste pipework should be 38mm UPVC: if your existing waste pipe is 32mm, be sure to get the right adapter joint.

Miscellaneous: Under this section come all the other bits and pieces needed to complete the job. Work out a comprehensive list once you've finalized the connection method.

Compression joints on supply pipes need jointing compound or PTFE tape, as do screw-on waste pipe and trap connections. Waste pipes are normally solvent jointed, for which you need the proper cement. If you add extensions of any length, you need wall brackets.

If you run a waste pipe through a wall, make sure you have the tools to make the

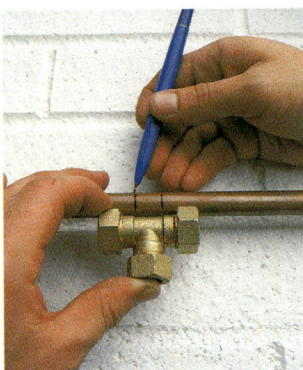

hole and the materials—sand and cement, mastic, plaster—to patch it afterwards. If the pipe connects to a gully, it should do so below the grid—buy a new plastic grid so that you can cut the appropriate holes.

Running new supply pipes

Once you've established your break-in point, the most daunting part of connecting the new supply pipe(s) is severing the existing ones. Yet it really is very easy: do the job methodically and there's little that can go wrong.

Start by draining down the pipes concerned; cut off their water supply, then open the taps at the ends of the runs to empty them completely.

The rising main (and all the cold pipes in a direct plumbing system) can be isolated by closing the main stop valve—unless, that is, you're lucky enough to find a local stop valve just before the break-in point.

In indirect systems, cold pipes other than the rising main are isolated by closing the valve on the main feed at the base of the storage tank. If there is no such valve, tie up the ball valve and turn on the taps to drain the tank.

In the absence of local valves, the supply to a hot pipe is effectively cut off by closing the valve on the cold feed to the hot water cylinder or heater. But in this case, be sure to turn off any heating systems first.

Note that if you decide to use automatic connectors there is no need either to cut off the supply or to drain down.

Breaking in: The procedure given here is for a standard compression-jointed tee connector.

Start by marking the length of the fitting against the pipe to be cut. Don't forget to allow for the fact that the pipe ends fit into it

by 20mm on each side.

Now cut the pipe at the first mark using a junior hacksaw. As you do so, make absolutely certain that the cut is square or the joint will be ruined.

Support the cut ends as best you can, then cut through the pipe at the second mark in exactly the same way. Follow by filing off the burred metal around the pipe ends; use a flat file for the outside edge and a round file on the inside. Connect the tee to the pipe run as shown in the sequence below.

Automatic connector: The instructions here are for a Thorsman T-plus. First make sure that the connector is the appropriate size for the pipe. Then clamp it over the break-in point using the Allen key provided.

Connect the rest of the new pipe run to the branch on the fitting and make sure the stop valve you have fitted is in the closed position—this is vital: the supply pipe still has water in it, and if you activate the cutter on the fitting before the branch is closed you'll have a flood on your hands.

To pierce the supply pipe, remove the plastic cover on the strike pin. Support the pipe run with pieces of wood, then hit the pin hard with a hammer. The charge inside the fitting will cut the hole, and the branch will be supplied with water.

Connecting the run

How you do this depends on how far the supply has to go and what fittings you use. If you are running hot and cold pipes, keep them together.

The conventional sequence after the tee is: section of pipe/stop valve/section of pipe/washing machine connector. All these can be compression-jointed in the sequences

1 *Mark the length of the tee fitting against the pipe. Cut squarely*

2 *Fit a compression tee joint then insert a branch pipe. Tighten cap nuts*

3 *Connect the branch pipework before using an automatic connector*

4 *Fit a stop valve before the flexible hose connector. Note the colour coding*

illustrated (below). Make absolutely certain, though, that you cut the ends of the pipe sections square or leaks are bound to occur. Secure the new runs where possible using plastic pipe clips pinned to the wall. A washing machine connector with built-in stop valve simplifies the run and is fitted in exactly the same way.

Extending the run: This is easily done, if necessary, using straight sections of pipe in conjunction with elbow fittings and bendable connectors. Follow the rules below to avoid problems.

• Keep hot and cold pipes running together, about 25mm apart. Choose a route that is unobtrusive—along the skirting or in the angle between wall and ceiling are two favourites.

• On no account let the run contain any sharp kinks or inverted U bends that could restrict the flow or create airlocks later.

• Secure straight runs of pipe every 1m using plastic clips.

• Proper planning will usually avoid the need to run pipe through a wall. If it is necessary hire a drill and follow the method described for the waste pipe (below). Run the pipe through plastic waste pipe.

Arranging drainage

If you are lucky, this is simple. But even adding a new waste pipe is not the daunting task it first appears.

If you are breaking into an existing waste pipe with an automatic connector, start by clamping the special fitting over the break-in point; tighten the four screws securely. Next, using the special tool provided with the fitting, bore a hole in the waste pipe through the hose outlet nozzle. Finally, screw the nozzle/valve assembly into the fitting until hand-tight.

Washing machine trap: measure this for size against the existing trap. Mark off on the new trap where it has to be cut.

> ### ★ WATCH POINT ★
>
> To ensure a square cut, wrap a piece of cardboard around the trap inlet and align it with the mark. Level it up all round, then use it as a template while you cut the pipe with a junior hacksaw.

Place a bowl under the old trap to catch the water in it. Unscrew the trap inlet and outlet joints, remove the old trap and fit the new one in its place. Wrap a few turns of PTFE tape around the joint threads to guard against leaks. Finally, screw on the hose outlet nozzle (note that this has no valve—the sink overflow will prevent siphonage).

New waste pipe: In this case you must have a standpipe. The standpipe should end at least 600mm above floor level; secure it to the wall behind where the washing machine is to go using the clips provided.

Plan the run of the new waste following the rules given in the diagram on page 58. The 38mm waste pipe can be solvent, compression or push-fit jointed. Clip it to the wall every 1m or so.

Waste pipe branch: Do this job with the 38mm tee fitting of your choice. Sever the existing pipe with a junior hacksaw, having measured and marked on it the exact size of the fitting. Use a cardboard template to keep the cuts square.

Gully connection: In this case the pipe will probably have to pass through an outside wall. On a masonry wall, mark the site of the hole and drill a pilot hole from inside

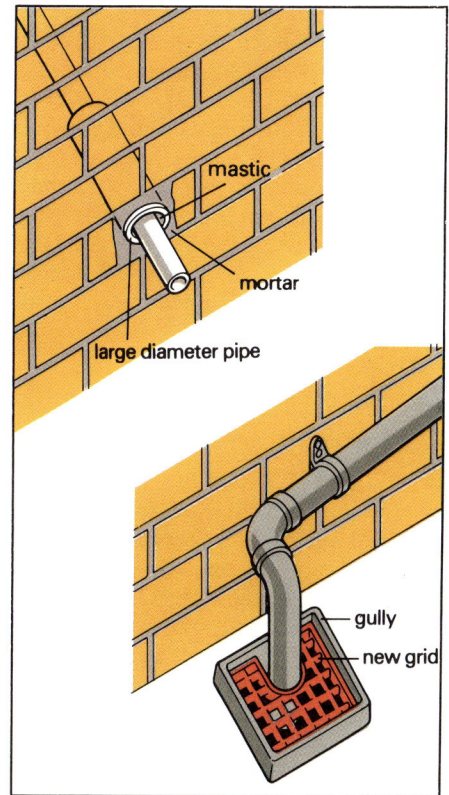

Top: where the pipe passes through a hole in the masonry, mortar in the larger pipe for it to run through and seal with mastic. Above: To connect to a gully, fit a new grid and lead the pipe through it, to 50mm below grid

using a masonry bit. Convert this into a hole just larger than the pipe diameter using a hammer and a long cold chisel. Or, alternatively, hire a heavy duty masonry bit or core drill from your local store and do the job with this.

Run the new pipe to the gully in the most convenient and straightforward way—but still following the rules on page 58—and

1 *Break-in connectors clamp over the pipe and a cutter is used to bore the hole*

2 *Washing machine traps replace the existing trap under the sink*

3 *Standpipe kits are simple to make up. Ensure that the pipe is at the right height*

4 *Chisel a hole in the wall to lead out the waste pipe. Work from both sides*

secure it at metre intervals.

When you reach the gully, replace the existing grid with your new plastic one, cut to accommodate the pipe. The pipe should terminate at least 50mm below the grid line but above the water line.

Make good the gaps in the wall around the pipe—with mortar on the outside, filler or plaster on the inside—but leave a gap of about 2mm all around the pipe. Caulk this afterwards with non-setting mastic—inside and out—so that the pipe can expand and contract without stress.

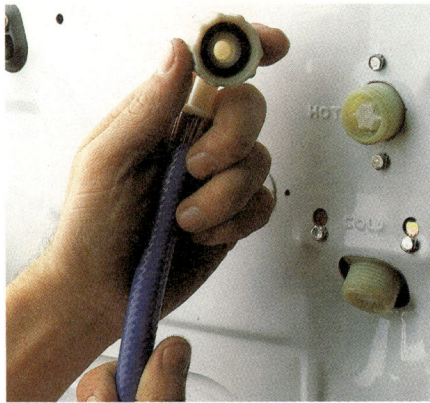

Connecting the machine

Always follow the manufacturer's instructions on this point. The washing machine hoses simply screw on to the relevant connectors—red for hot, blue for cold. The drain hose can be slid onto the outlet nozzle or simply slipped into the top of the stand-pipe (don't fasten it), depending on your drainage connection.

Joining pipes

There are many different ways to join two sections of pipe. The only ones which are important here are compression and push-fit for water pipe, and push-fit and solvent-welded for plastic drainage pipe.

Whatever sort of joint you use, make sure that the pipe is cut accurately square and to length. Remove any burrs.

Always ensure that the diameter of the joint matches that of the pipe. This may not be immediately obvious. In the case of the water pipes, there should be no problem—only 15mm pipe is concerned. And with

Below: instead of plugging it in, run the machine off its own fused connection unit

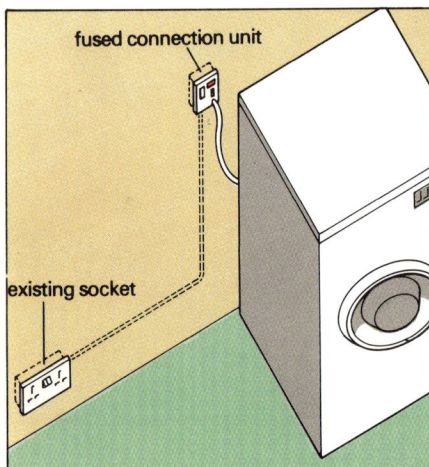

1 *Connect the hoses to the supply and outlet. Note the colour code*

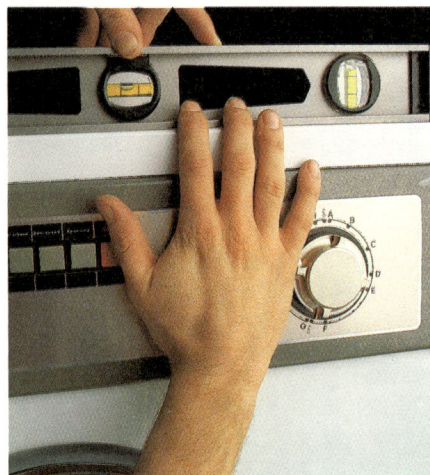

2 *Make sure the machine is properly levelled before use*

compression and push-fit joints, you can even connect to the older Imperial ½in. pipe (which you may well have) without using an adapter.

The waste pipes may not be so simple. There are two standard sizes—38mm and 32mm. But even within the same size, different makers' pipes are not necessary interchangeable. It may well be essential to use the same brand for the extension.

Compression joints

Remove the capnut from the joint and slip it onto the pipe. The seal is made by the small ring under the capnut—called an olive. Slip this over the end of the pipe. If you are fitting to plastic pipe, you also need a metal insert to stiffen the end.

Push the end of the pipe into the joint until it butts firmly against the end of the recess. Wrap a little PTFE tape around the threads of the joint then screw up the

capnut finger-tight, making sure that the pipe doesn't pull out of place.

Now grip the body of the fitting with a spanner or adjustable wrench, and place another on the nut. Tighten it to compress the olive—but restrict this to one or two turns at most to avoid damage. Some makers provide specific instructions on this point—follow them.

For a better seal—particularly when joining to Imperial pipes—apply pipe jointing compound to the mating surfaces of the joint and pipe.

Push-fit joints

Push-fit joints for water pipe are somewhat similar to compression joints—except that the seal is made by a rubber ring and you can do the nut up by hand. As well as the seal, there is an additional metal grab-ring which stops the pipe pulling out.

Stiffen plastic pipe with a metal insert. With either plastic or copper, smear a little silicone lubricant on the end and insert it through the seal and grab-ring. Tighten the capnut firmly by hand.

Some waste pipe joints are similar—although no insert or grab-ring is used. With others, there is no capnut and the end of the pipe is simply pushed firmly into place. It is important that these joints are not strained.

Solvent-weld joints

Unlike the other joints described, these are permanent—so you must get them right first time. Dry assemble first and mark both joint and pipe so you can align them properly once the joint is made.

Clean the mating areas thoroughly—there is a proper cleaner but methylated spirits is adequate. Then spread solvent cement on the mating area of the end of the pipe, aiming for a thin, even coating. Immediately push the pipe home in the joint and hold in alignment for 15 seconds.

★ WATCH POINT ★

One final point that people often neglect is that the machine must be level, or the balance of the drum could be upset. Check this with a spirit level after you have connected the machine and slid it into its final position.

CHAPTER 4

PLUMBING EXTRAS

Water filters, sink sprays or an extra WC all sound like luxury items but they are relatively inexpensive and can help to make your home much more pleasant and efficient. Installation is no problem—all the work can be carried out using standard DIY plumbing fittings.

BUILDING A VANITY UNIT

The road to work isn't the only area that's notorious for early morning traffic jams: even before you set off you may find the bathroom prone to congestion, as members of the family queue up for a turn at the single washbasin.

One obvious way to ease the flow is to create a second bathroom. But space—or finance—may not permit this. A second washbasin is a more modest, but just as practical and convenient way to get rid of the usual bottleneck. And if you install it in a spare bedroom, you'll find it a boon whenever you have visitors to stay.

A basin simply fixed to the wall can look rather clinical in a bedroom; but install it in its own vanity cupboard—and mount a mirror, shaver socket or light on the wall above—and you've a versatile unit for making-up, hair styling, or simply washing. Not only will the unit conceal all the pipework under the basin but it will also provide handy storage for toiletries and cosmetics.

Plumbing-in the vanity basin isn't complicated, especially if you use plastic push-fit connectors and flexible plastic pipe for the long runs often necessary to reach the bedroom. Plumbing-out the waste run is also straightforward, using larger push-fit connectors and plastic pipe.

Choosing a vanity unit

Ready-made vanity units are available from the major sanitaryware manufacturers, designed to take their range of basins. Different styles are made to fit into various locations: in an alcove, within a run of cupboards, or as a freestanding unit against a wall. They're commonly made from easy-to-clean white, coloured or patterned melamine-faced chipboard. You can choose from models with plain, patterned or louvre doors with various knobs and handles, and there's a choice of tops with rounded front edges and a coloured or textured finish. Some types resemble a desk, with two side cupboards so you can sit comfortably facing the basin. Most units include a recessed plinth or toe hole.

To fit the basin you have to cut a hole in the post-formed top (some units come ready-cut) using a sealing gasket as a template, then drop in the basin and then make its supply and plumbing connections, which are exactly the same as an ordinary basin.

Ready made vanity unit cabinets come in a range of styles and colours. The basin may overhang, be inset into the top, or form the whole of the top. Taps may be separate or mixer—and some have pop-up wastes

Basins

The choice of basins is vast. They're usually made from vitreous china, plastic or enamelled pressed steel, in a range of colours so wide that you're sure to find one to complement your bedroom decor.

Basin shapes, too, are numerous: you can choose from rectangular, oval, square, round, deep or shallow dishes. Some basins are designed to project beyond the face of the cupboard, and you can even buy a vanity basin that spans the entire width of its cupboard, replacing the usual post-formed top.

Taps

Again, there's a large selection of taps to choose from, including conventional basin-pillar types, or two- or three-hole mixer units, and the newer one-hole or 'mono-block' mixers, which combine hot and cold taps and spout in a single body (see diagram). Whichever you choose, the installation is similar, although a one-hole mixer requires a special reducing connector to attach it to the water supply.

When choosing a basin and taps, look at the types of waste outlets also; you can have the normal outlet with plug and chain, or a pop-up waste that's fitted to the mixer tap unit or set in a separate central hole in the basin, between the taps.

Consult manufacturer's leaflets and visit

showrooms before choosing—you may be able to mix and match basins, taps and cupboards from different manufacturers if you can't find exactly what you want in a particular range.

The plumbing connections

Connecting the basin to the water supply and waste system is straightforward (see pages 67–68): you take a branch out of the water supply pipes with a tee connector and run in a new waste pipe. But there are some plumbing regulations you should be aware of (see below).

There's no need to bother with copper pipes and compression or soldered capillary fittings—all the connections can be made using plastic pipe and push-fit connectors, without the need for any specialist tools. You can buy all the plumbing components individually from DIY stores, or in kits containing all you'll need to complete the job. Coils of flexible polybutylene pipe are sold in large quantities to cope with long runs.

Your main consideration is how and where to route the new pipes: the bedroom —or other room where you plan to install the vanity unit—may be some distance from the connection point, and you'll have to plan carefully to avoid unnecessarily complex runs. You'll also have to plan how best to hide the pipes.

Cold water supply

Where you obtain your new water supply depends on your plumbing system (see diagram). In a modern 'indirect' system, you can take the new cold water supply for an upstairs vanity basin either from the cold water storage cistern in the roof space via a tank connector, or tee into the bathroom cold water pipe feeding the cold tap. On no account must you take the supply from the cold water feed to the hot water cylinder.

Hot water supply

The new hot water supply can be a branch taken from the hot water supply pipe to the bathroom tap, or from the vent pipe from the hot water cylinder *below* the base of the cold water cistern.

If the vanity basin is to be on the ground floor you can take its cold water supply from the rising main or a branch from it, and its hot supply from the hot water pipe supplying the hot tap in the kitchen. Before you

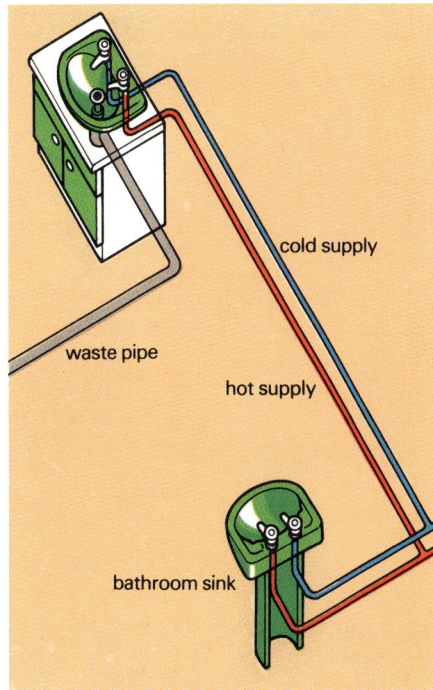

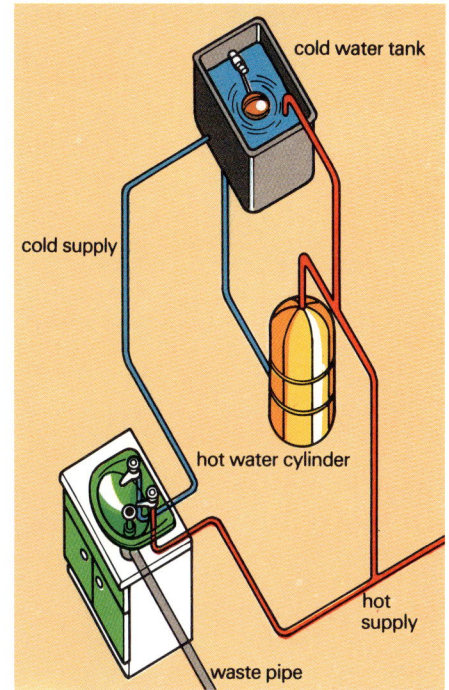

Arrange the supply for a vanity unit to keep the pipe route simple. The two alternatives are that you can either tee into the hot and cold supplies of an existing basin (above), in the bathroom for example, or take the cold supply from the storage tank in the loft and the hot supply from the vent pipe from the hot water cylinder (above right)

go ahead, though, consult your local water authority: some forbid the householder from carrying out any alterations of this type to the rising main in case it becomes contaminated.

The waste run

The way you plumb-out the new waste pipe again depends on your type of system (see diagram). In the old 'two-pipe' set up (with two waste stacks: one for waste water; one for soil waste from the WC) the plumbing fittings discharge either directly into the stack, into a hopper (on first floors), or into a yard gulley (at ground level). The new waste pipe can be run into either of these, or you can tap into the waste pipe from another plumbing fitting, using a 'swept' waste connector.

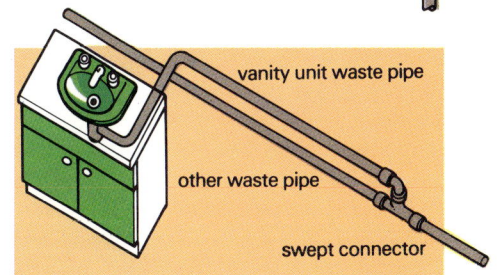

In the newer 'single stack' system (one stack into which all the plumbing fittings and soil waste pipes discharge) you'll have to connect directly into the stack using a 'boss' connector. Never, with a single stack system, break into an existing waste pipe.

The waste from the vanity unit can be run to a hopper, gully, stack or existing waste pipe, but do not break into an existing waste pipe if your house has single stack drainage. Keep all waste pipe runs as simple as possible

What to buy

Apart from the basin, cupboard unit and suitable taps, you'll need two 15mm equal push-fit tees (22mm reducing to 15mm if the supply pipe is the old type) to branch out of the water supply pipes, sufficient plastic pipe to reach the vanity unit (plus metal supporting inserts wherever there's a connection). If you're connecting into the cold water storage tank you'll need a 15mm push-fit tank connector.

At the basin end of the run you'll need two 15mm push-fit straight connectors to join the plastic pipe to two lengths of 15mm hand-bendable copper pipe, which is used to connect the supply to push-fit/compression tap connectors. These in turn are attached to the tap tails (you'll also need two special reducing connectors to connect the small-diameter 10mm or 12mm bendable copper tails of a one-hole mixer to the 15mm supply pipe).

To connect the waste outlet of the basin to the hopper, gully or stack you'll need a 75mm bottle trap and sufficient 32mm or 38mm plastic waste pipe (depending on the length of run) and push-fit connectors to complete the run. If you're connecting into an existing waste pipe you'll need a 32mm or 38mm push-fit swept connector.

Assembling the unit and basin

Once you've run in the new water supply and waste pipes, all you have to do is locate the basin in its vanity unit, connect the taps and waste trap, then make the final connections into the water supply.

Most vanity units are easy to assemble with KD (knock-down) or other proprietary joints—some even come ready assembled.

If you have to cut a hole in the unit top, either mark out its position using the sealing gasket supplied with the basin as a template, or make a template of the basin from newspaper. Cut out the basin profile using a jigsaw.

Position the unit and secure to the floor and wall using joint blocks. Before you place the basin in its cut-out, it is easiest to fit the tap. Assembly varies from make to make but basically you slot the taps and their washers into the holes in the basin (remember, it's usual to have cold on the right; hot on the left), slip a washer onto their threaded tails and secure with back nuts. Some taps should be bedded on a ring

1 *Assembling vanity units rarely presents problems but make sure that all the necessary fixings are supplied*

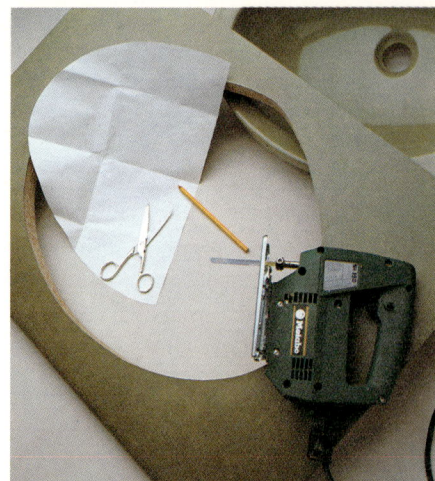

2 *Cut a hole in the vanity unit using the template supplied or make your own from a piece of card*

4 *Mixer taps with 10mm or 12mm tails need reducing adaptors to match the 15mm supply*

5 *Remove the nut on the 15mm side of each adaptor and screw on a plastic tap connector*

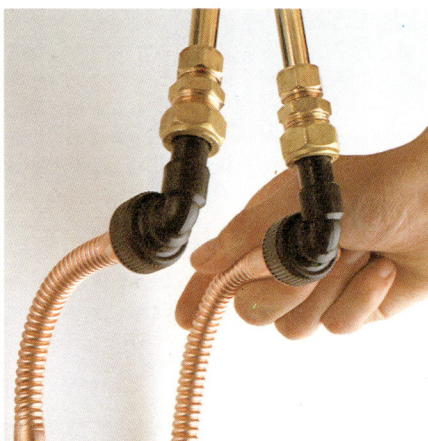

7 *Lubricate the ends of the flexible copper pipes and push them into the tap connectors*

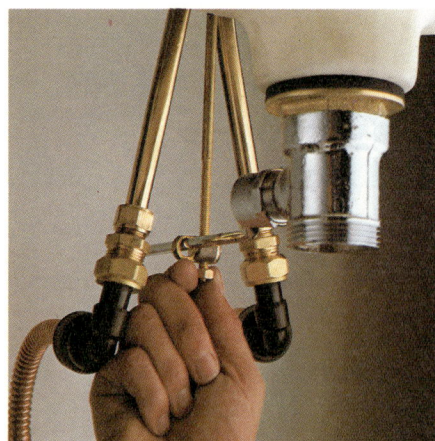

8 *If your basin has a pop-up waste, make the necessary connections and then test it*

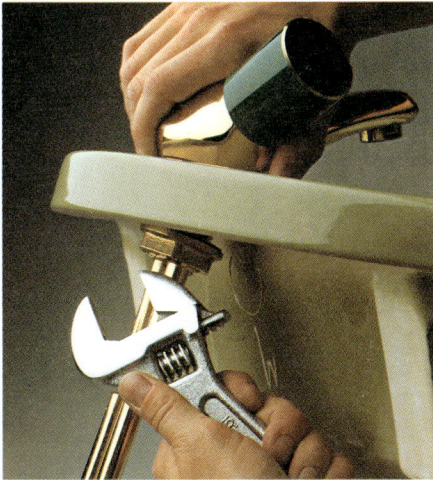

3 *Fit the tap before installing the basin —don't forget any washers under the backnut before you tighten up*

6 *Fit the sealing gasket around the cut-out and then lower the basin into place. Check for gaps*

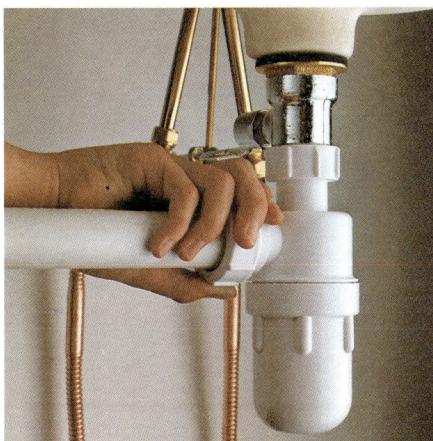

9 *Complete your work at the basin itself by attaching the 75mm bottle trap to the outlet*

of mastic instead of the washer.

Next, screw on one tap connector per tap and neatly insert the supply pipe to the push-fit end.

Remember to deburr and lubricate the ends of the copper and plastic pipes before insertion, and lubricate inside the connectors, too. You'll also have to fit a short section of a special metal supporting insert into the end of the plastic pipe to stop it collapsing under pressure.

If you're fitting a three-hole mixer unit, insert the outlet spout and its packing washers first and secure with the brass nut provided in the central hole of the basin.

Next, fit the water inlet assembly from underneath the basin in the remaining two holes and secure at the top with brass nuts; tighten with an adjustable spanner. Slot the flange and headwork onto the protruding shanks of the inlet assembly and secure by tightening the retaining nut. Complete the tap head assembly by fitting the coloured indicators and decorative tap heads (cold on the right, hot on the left).

If the assembly includes a pop-up waste, first fit the waste outlet on its rubber washer, turn the basin bottom-up and fit the pop-up waste control rods, which may need shortening to give sufficient clearance inside the vanity unit.

If you're fitting a one-hole mixer tap, fit the unit in the same way as a conventional tap in the basin's single central hole, then attach a special reducing connector to the twin 10mm or 12mm copper bendable inlets that protrude from the underside of the taps. These are secured to the mixer and supply pipes with compression joints.

Again, you can fit pliable copper pipe between the tap connection and the supply pipe (see above).

Before you drop the basin into the cut-out

in the vanity unit top, apply a strip of mastic around the opening to ensure a watertight seal, or fit the special sealing gasket supplied with the basin and unit. Take care to achieve a good seal at this stage. If you don't, you'll only have to undo much of your hard work later.

Lower the basin into place and secure it (if necessary) to the underside of the unit top. Push the pop-up waste plug into the outlet and test its operation dry.

Attach the 75mm bottle trap to the screw-threaded shank of the outlet and connect to the waste pipe.

Planning waste runs

A waste pipe is more difficult to conceal than supply pipe, so you may have to be content with running it along the skirting and boxing it in later—you wouldn't be able to cut sufficiently large holes for it in the joists. You may be lucky enough to be able to take the pipe from the trap straight out through the wall to the drainage point: cut a hole for it from each side with a club hammer and long cold chisel.

If the run is 1.7m long or less, use 32mm pipe; if it's longer—up to a maximum of 2.3m—use 38mm pipe. The run must slope gradually towards the drainage point; about 20mm per metre for a 2.3m run is usually adequate.

The push-fit connectors have no internal grab-rings (unlike those for water pipes) so you have to be sure they aren't placed under any strain, which could result in leaks. Clip the pipe to the wall at a minimum of one metre intervals.

If the pipe is to be connected into an existing waste pipe nearby you'll need to cut out a section of the pipe and fit a push-fit swept connector, which will allow a gradual flow from the branch into the main pipe. Two fittings sharing one waste pipe often results in siphonage problems however, so this is not an ideal arrangement.

Connecting to the water supply

When you've run in the water and waste pipes, and have assembled and connected the vanity unit, all that's left is to join into the water supply.

At this stage you should have your vanity unit assembled and connected to the new pipe runs. To complete the circuit you have to tee into the water supply pipes or join into the cold water cistern.

Before you can cut into the water supply you'll have to drain the relevant pipes or the cistern. If you're lucky, there may be local stopvalves with which you can isolate the pipes and drain them by turning on the taps they feed.

If there aren't any stopvalves, isolate the cold pipes by turning off the main stopvalve on the rising main, which prevents any more water entering the house (direct system), or turn off the cold water stopvalve at the base of the cold water cistern (indirect system), then turn on the taps to drain the pipe fully.

To drain the hot pipe turn off the stopvalve on the supply pipe entering the base of the hot water cylinder (although the supply will normally be shut off when the cold water supply from the mains or tank is shut off, or the tank is drained), then turn on the hot tap.

If you're fitting a tank connector, you'll have to drain the cold water cistern. To do this, turn off the feed to the tank and open the bathroom cold tap.

Once you've drained the pipes mark the positions of the push-fit tees at the break-in point and cut out the waste section using a junior hacksaw. Deburr and lubricate the cut ends, then push on the connectors, making sure they're fully seated inside.

The open ends of the new hot and cold pipes running to the vanity unit should be positioned so that you simply have to insert them in the branches of the tees.

To fit a tank connector, mark and cut a hole in the side of the drained tank about 25mm from its base, using a special tank cutting accessory for an electric drill. Dismantle the tank connector and reassemble it from both sides of the hole. Insert the end of the plastic cold water supply pipe into the push-fit connector.

Once you've made the final connection you can restore the water supply and check the system for leaks. When checking, get an assistant to control the stopcock so the supply can be turned off quickly if necessary.

10 *Mark and cut a section out of the existing supply pipes and fit push-fit tee joints. Deburr and lubricate the cut end before fitting*

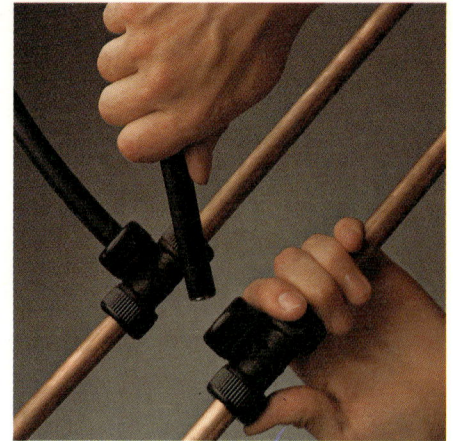

11 *Insert a metal support sleeve in the end of each flexible pipe, apply some lubricant and then push the pipes into the connectors*

12 *If you are taking the cold supply direct from the storage tank in the loft, buy a tank connector and a tank cutter or hole saw*

★ WATCH POINT ★

Release the supply pipes from their retaining clips at each side of the connection point to give enough play to push on the tee.

There is a large variety of basins and vanity units to choose from. Most are freestanding units with matching cabinets and basins. Before you commit yourself check that the basin is large enough for your purposes and that the cabinet has enough storage space for all your toiletries. Carefully measure the space you have available before purchase

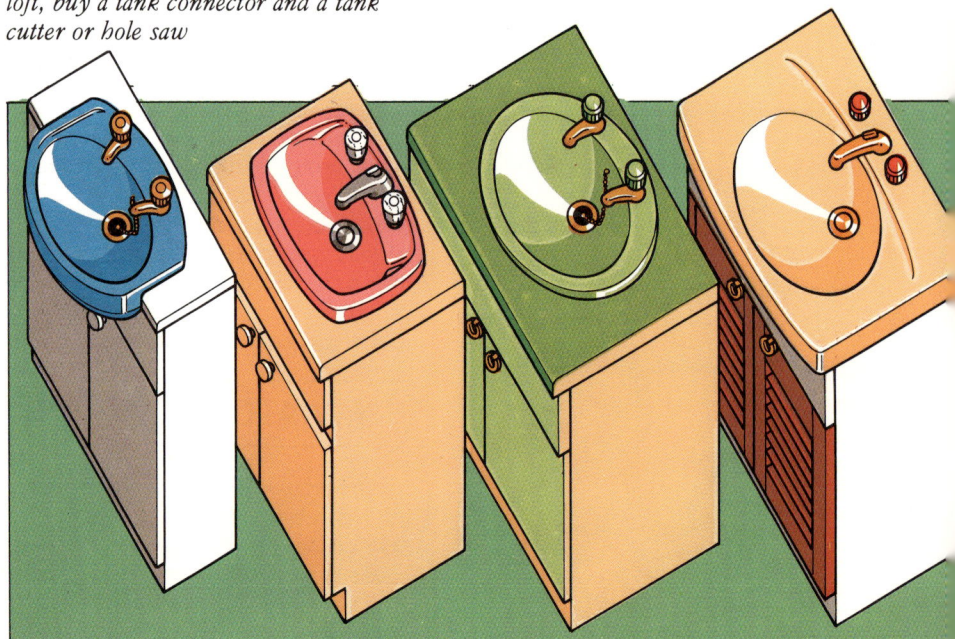

INSTALLING AN EXTRA WC

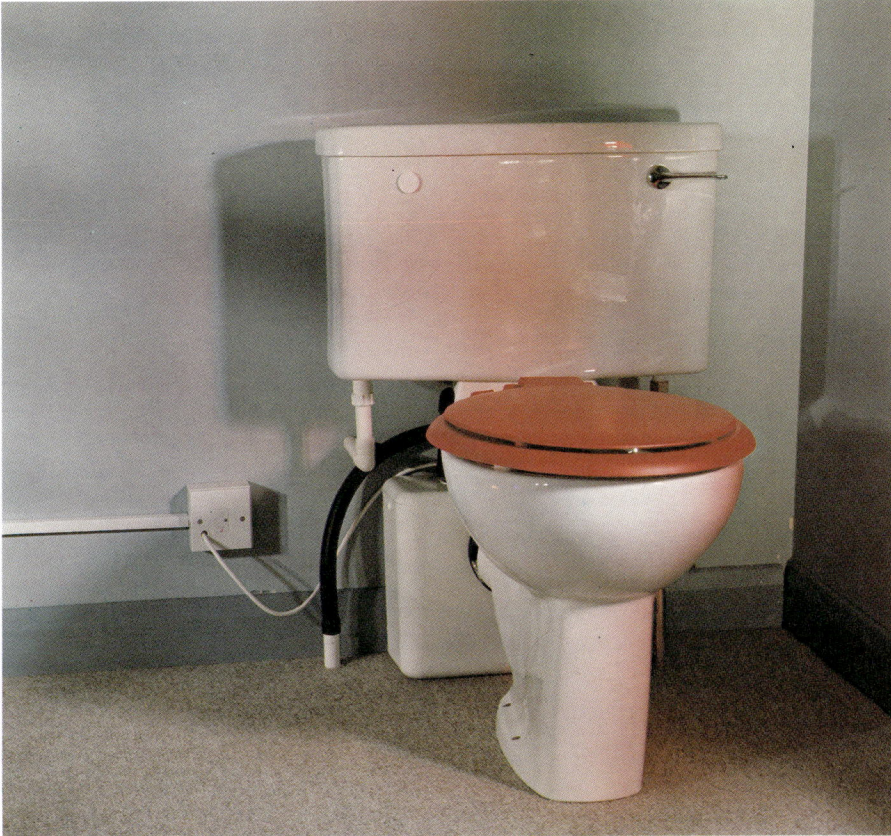

An extra WC can be a real asset in a home if, for example, you have an elderly or disabled person living with you. But it isn't always easy to install a new WC exactly where you want it because of the need to accommodate a bulky 100mm diameter waste pipe, which links the pan with the soil stack, and so the drain. In addition conventional waste discharge from a WC works on a gravity principle, which limits the length of the waste pipe (6m maximum) and requires a positive 'fall' to ensure that the waste drains away efficiently.

It is now possible to buy a device that eliminates the need for a large-diameter waste pipe with its limited length and steep 'fall'. Basically, it's a compact pump and shredder unit, which connects directly to the back of the WC pan and liquidizes solid waste before discharging it through small bore copper or solvent-weld PVC pipe to the soil stack. The pumped discharge allows a much longer pipe run between the amount of fall needed. However, the unit is only

meant for those locations where conventional drainage is impossible. Because of the special nature of the system you must have planning approval from your local authority before you begin the installation.

Operation of the device is automatic following the flushing of the WC. As the waste flows into the unit, it triggers a pressure switch, which operates an electric motor to drive the stainless steel shredder blades and pump. A fine mesh filter limits particle size before the slurry is pumped away through the outlet.

The discharge pipework needs a minimum bore of 18mm, which equates with the kind of pipework used in central heating systems. This makes it suitable for running floor/ceiling voids, hollow partition walls or along skirtings.

The system offers considerable flexibility in siting of the WC, making it ideal for garage or loft conversions, extensions and so on. It is advisable to site your new WC as close to the existing soil stack as possible to

minimize the amount of disruption to your home during the installation.

Planning considerations

The most important point to remember is that the installation of a new WC, or any plumbing work that involves modifying or adding to the building's existing waste system, must have the approval of your local building inspector. So, consult him with your plans before you begin work; he will probably want to inspect the finished installation.

You must have access to a suitable electrical circuit to provide power for the unit and a suitable cold water supply to feed the WC cistern.

The unit can be positioned up to 30m from the soil stack, but ideally it should be near an outside wall, through which the cistern's overflow pipe and the unit's vent pipe can be passed if that is acceptable to your local authority. Plan the pipe run carefully, keeping it as straight as possible and bearing in mind the need for a slight fall of 5mm in every metre.

The unit will pump the waste vertically, if required, but the maximum distance is limited to 2m—an important consideration if you're installing it in a basement.

Ample ventilation is essential, and if there is no openable window nearby, you'll have to fit an extractor fan. Also, remember that a room containing a WC must not open directly on to a living room or a kitchen; there must be a ventilated lobby in between.

Tools and materials

The exact tools you require will vary from one installation to another and with the pipe materials chosen. However, the following are essential: spirit level, steel measuring tape, pencil, electric drill, masonry and wood bits, bradawl, screwdriver, junior hacksaw, flat and half-round files, wire wool, two adjustable wrenches, long cold chisel, club hammer, wire strippers or a sharp knife. A small electrical screwdriver is also necessary.

If you plan to use copper pipe with capil-

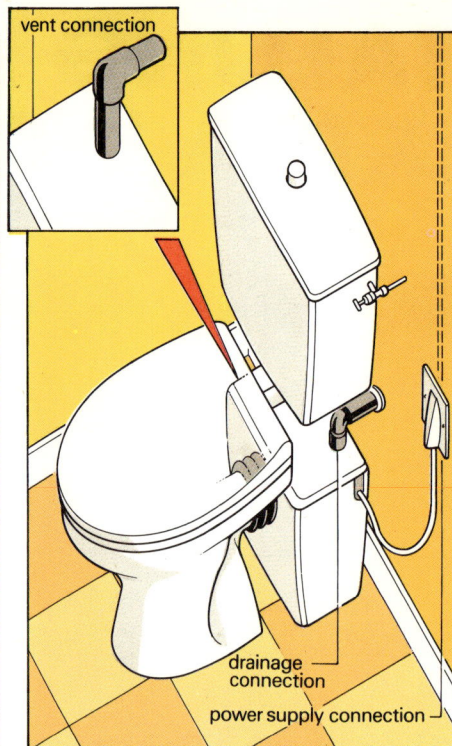

vent connection

drainage
connection

power supply connection

An extra WC can be installed in several locations around the house. If you have an elderly or disabled person living with you an added WC in a 'granny' flat is a great help. If you are planning a loft or base-ment conversion, an extra WC is well worth considering. The pump in the unit enables you to locate a WC below your existing drainage run provided that the distance is less than 2m. Locate the new WC with care to allow for as little structural change as possible

lary fittings, you'll also need a blow torch, flux, solder (if you're using end-feed fittings), protective gloves and a flame proof mat (to protect skirtings). If you're using plastic solvent-weld pipe and fittings, you'll need a tin of cleaning fluid and a tin of solvent cement.

For the actual plumbing, you'll need sufficient lengths of copper or PVC pipe and the appropriate fittings to make the vent and waste runs from the unit, the feed pipe to the cistern and to take the overflow pipe from the cistern.

Jubilee clips are needed to connect the vent and waste outlets of the unit to the appropriate pipe runs and to secure the flexible coupler to the pan outlet. If there is no convenient boss branch on the soil stack to connect into, you'll want a suitable strap or saddle boss to make the connection.

The unit needs a 3 amp fused connection unit with front flex outlet suitable for use in bathrooms and this, in turn, can be fed from

In most situations local authorities insist that the unit be vented to an external wall or into the existing soil stack. In some cases it is possible to vent the unit internally but the air inlet cowl must be above the height of the WC pan

a lighting circuit (with 1.5mm² cable) or a ring main (with a 2.5mm² cable). If you're not taking the power directly from the back of a socket or loop-in ceiling rose, you'll also need a three-terminal junction box of the appropriate rating.

Other items you may need are: a length of conduit, cable clips, wall plugs, a small bag of ready-mixed mortar and some all-purpose filler for making good.

Installing the unit

The shredder unit is freestanding, so there's no need to make any mounting holes; simply position it where you want the WC and connect up the outlet pipes.

Stand the shredder unit in the appropriate position close to the wall, allowing enough clearance so that it won't be wedged in place by the pan when that is installed.

Connect the vent pipe first; exactly what you do here depends on the requirements of your local authority. Generally, it should be taken through an outside wall or connected to the soil stack so that it can be safely vented to the atmosphere.

You can use either copper or solvent-

weld PVC pipe for the vent. Cut a short length and push this into the moulded vent outlet on the top of the unit, securing it with a jubilee clip. If the unit is against an outside wall and you are not required to connect the vent to the soil stack, you can simply take the pipe back through that wall: cut a suitable hole with a cold chisel and club hammer, keeping it as small as possible to reduce the amount of making good needed later. Run a length of pipe through the hole and connect it to the short length already attached to the vent outlet with an elbow fitting. On the other side of the wall you may be required to run the pipe to above eaves level, so fit another elbow and run the pipe vertically upwards. Clip it at 1.2m intervals and fit the cowl supplied with the unit to the top.

If there's no outside wall nearby, or if your local authority requires you to connect the vent pipe to the soil stack, run the vent pipe alongside the waste pipe.

The unit's waste outlet is an elbow fitting on the top, which may be turned as necessary to face the direction of the waste pipe run. Adjust its position and then take the flexible pipe supplied and push one end over the elbow's spigot, securing it with a jubilee clip. The other end of this pipe should be taken to the rigid waste pipe run, making sure that any bend in it has as large a radius as possible to prevent it becoming kinked and causing a blockage. Push the end of the flexible pipe over the end of the waste pipe and secure it with a jubilee clip.

Make the pipe run to the soil stack, keeping it as straight as possible, with any bend being of large radius. Keep the number of fittings to the minimum and clip the pipe to the wall at 1.5m intervals so that it's supported adequately.

When running pipes beneath the floor,

<table>
<tr><td>★ WATCH POINT ★</td></tr>
<tr><td>Lubricating the lip of the coupler will help in fitting it to the pan outlet. Use petroleum jelly or a little washing up liquid to make the job easier.</td></tr>
</table>

clip them to the sides of the joists or to battens fitted between the joists. If they need to cross the joists, this should be done at right angles to the joist run. Cut notches for the pipes in the tops of the joists so that they're positioned beneath the centres of the floorboards above; this will reduce the likelihood of them being accidentally pierced by a nail when the boards are replaced. Don't cut the notches any larger than they need be, otherwise you run the risk of weakening the joists.

Allow a fall on the pipe of 5mm in every metre, and after about 13.5m you'd be wise to connect into pipework of larger bore (about 40mm is adequate), which will reduce the load on the pump. If you have laid out the pipe run so that the unit pumps vertically upwards at first, the discharge pipe should be connected to a pipe with a minimum bore of 30mm and a positive fall immediately after the vertical run. This will ensure that the waste flows away freely and that there's no possibility of backflow or back siphonage.

Run the waste pipe as close as possible to the soil stack before breaking through the outside wall (assuming, of course, that the stack is not run in a duct inside the house). Cut a short length of pipe to run through the wall and connect it to the pipe from the unit with an elbow fitting. Add another elbow and straight length of pipe to link to the soil stack itself. Run the PVC vent pipe to an external wall or the soil stack using the same method as you have for the waste pipe.

When all the pipework has been completed, the pan and cistern can be installed.

You may be faced with connecting into either a plastic or cast-iron soil stack, and the former is much easier than the latter due to the 'workability' of the material.

If you are lucky, there may be a convenient boss branch with a spare entry point on a plastic stack, but even if there isn't, connection is still straightforward with a self-locking boss connection.

If there is a spare entry, simply open it up with a hole saw—or drill a series of holes round the edge and cut out the blanking piece with a padsaw—and fit a special connector. The waste branch must be solvent-welded to the connector.

A self-locking boss is fitted by cutting a hole in the side of the stack, inserting one half of the boss and screwing the outer half to it. The waste branch is fitted as before.

The methods of breaking into a cast iron stack are similar to those used with plastic stacks. However, since the material is more difficult to cut, it may be easier to remove a

1 *Position the shredder unit. Attach the internal vent or mark the external vent position; clip into place*

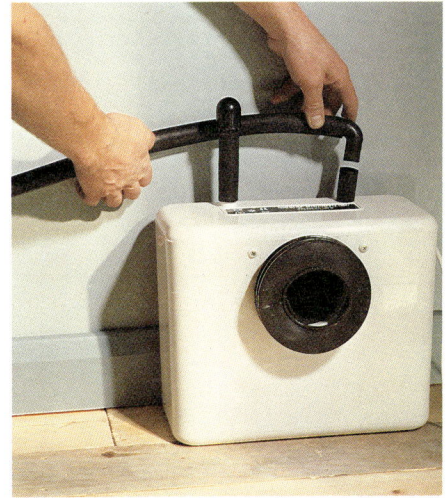

2 *Attach the flexible pipe to the waste outlet fitting of the shredder unit. Secure it with a jubilee clip*

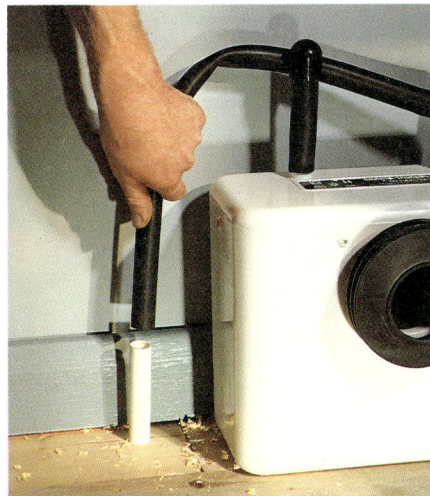

3 *Fix the other end of the flexible pipe to the end of the rigid waste pipe run. Keep curves to a minimum*

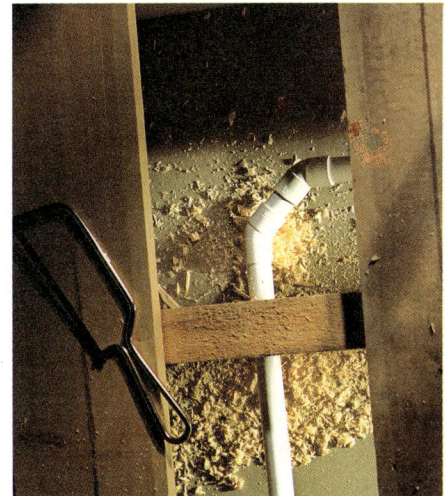

4 *Run the rigid waste pipe under the floor boards to the soil stack, clipping the pipe to the joists where necessary*

complete section and replace it with a length of plastic pipe. It is probably better to replace the entire top section of the stack as

you will have only one connection to make. Remove a convenient section and use it to mark the length of the new plastic pipe, remembering to subtract the length of any new connectors. Use a special fitting, of which there are many types, to connect the

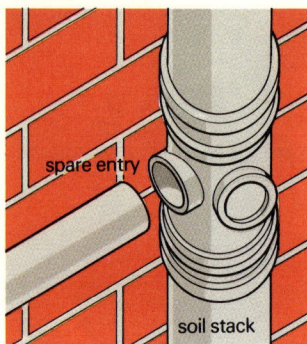

Join the plastic waste pipe into the stack by taking out a spare entry blank on a boss branch. Solvent weld the pipe

Use a self-locking boss if you don't have a convenient boss branch on the stack, or if all the entry holes are in use.

5 *Connect the flexible coupler to the WC pan and partly tighten the jubilee clip holding it in place*

6 *Check the pan for level using a spirit level. Secure it tightly to the floor using brass screws*

★ WATCH POINT ★

If you are using a drill and padsaw to remove a disc of material from the stack, drill a hole in the centre of the disc first. Bend a piece of stiff wire so that one end is hooked and the other is formed into a hand loop. Insert the hook end through the hole with the hook uppermost and hold the loop while you cut between the drill holes. This will prevent the disc falling into the stack, where it could possibly cause a blockage.

---

old and new stacks. Break the plastic waste pipe into the stack in the same way as described before.

Installing the WC pan and cistern

Once you've installed the shredder unit, you can position the WC pan and cistern and make the plumbing connections.

Set the pan in front of the shredder unit and slide it back so that you can pull the unit's flexible coupler over the pan outlet. Make sure that it isn't distorted or under any strain—positioning the pan square on to the unit will help here.

When the pan is sited properly, mark the positions of the mounting screw pilot holes on the floor by pushing a bradawl through the screw holes in the base of the pan. If you haven't a bradawl, or if the blade isn't long enough, you can use a long nail or even a skewer to mark the position.

Remove the pan temporarily and drill the pilot holes; if the floor is solid concrete, drill the holes with a masonry bit and fit plastic wall plugs.

Replace the pan and draw on the shredder unit's flexible coupler. Screw the pan to the floor, using brass screws with lead washers beneath their heads—these will prevent you overtightening the screws and cracking the pan. Before you fully tighten the screws, lay a spirit level across the top of the bowl—from side to side and from back to front—to check that it's sitting perfectly level. If it isn't, insert some slivers of wood below the base of the pan to level it. Fully tighten the screws then clip the flexible coupler to the pan outlet.

Next, offer up the cistern to the wall, following the manufacturer's advice on its height above the pan. If none is given, fit the cistern at about waist height so that the flush handle will come comfortably to hand. Mark the position of the cistern on the wall, using a spirit level for accuracy.

Remove the cistern and position its supporting brackets so that they coincide with the marks you've just made. Mark the screw holes then drill and plug them; screw the brackets to the wall. Place the cistern on its brackets—checking once more that it's level—and mark the positions of the upper screw holes; drill and plug these before screwing the cistern to the wall.

Fit the fall pipe between the cistern and the pan, checking that the rubber seals are properly located; you may have to trim one or both ends of the pipe for it to fit, so offer it up first just in case.

Assemble the internal flushing mechanism, according to the manufacturer's instructions, then attach the float arm assembly. Complete the installation by fitting the flush handle and linkage.

You are now ready to connect the cistern to the water supply, and you can make the pipe run in 15mm copper or plastic pipe.

Continue running the pipe back to the point where you intend breaking into the existing cold water supply, clipping at 1.2m intervals and using as few bends and fittings as possible.

When you reach the point where you intend to break into the supply, cut the last piece of pipe so that it's a little too long. Turn off the main stopcock or tie up the cold water storage cistern float arm and drain down the system by opening all the cold taps (except the one over the kitchen sink, which is connected directly to the mains supply).

When the system has been drained, offer up a tee fitting to the supply pipe and mark the latter for cutting, using the depth stop mark on the body of the fitting as a guide.

With the supply pipe cut, spring the tee fitting into place and then mark the new pipe to length to fit neatly in the branch of the tee. Before restoring the water supply, connect the plastic overflow pipe to the cistern and run it to an outside wall.

Allow the system to fill and check for leaks at all the pipe joints.

Powering the unit

To complete the shredder unit installation, you need to connect the unit itself to the power supply. You can take power either from a nearby lighting circuit or from the ring main in the house.

Whichever method of powering the shredder unit you choose, you **must** remove the fuse controlling that circuit from the main consumer unit before you start work.

First mark the position of the connector unit's box on the wall and cut a recess for the box with a bolster chisel. Screw the box to its recess, having removed the

7 *Fit the cistern making sure that the fall pipe and rubber seals are properly located and do not leak*

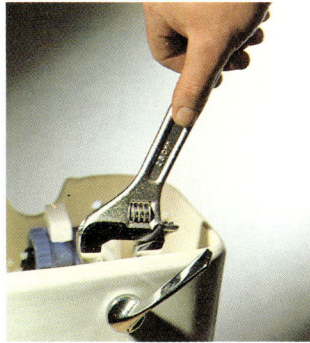

8 *Assemble the internal flushing mechanism. Attach the float arm and fit handle. Adjust the arm*

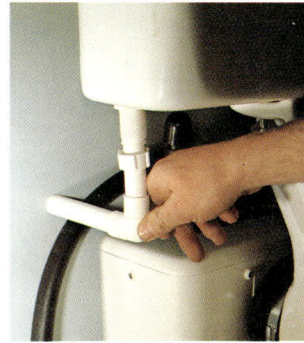

9 *Connect the water supply pipe to the cistern; carry it to the existing water supply. Connect up the overflow*

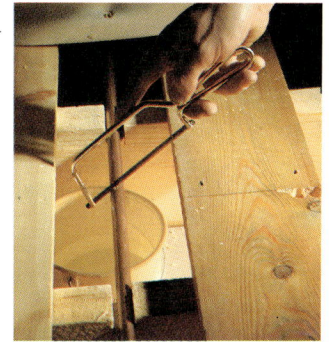

10 *With the water turned off, mark, and then cut, the pipe you'll tee into. Try to exit at right angles*

appropriate knockout in its side to allow cable entry. Then, chop out a cable chase from that knockout to either the ceiling or the floor, depending on where the cable is to run from. Although you can clip the cable into the chase, it's safer to run it in a length of conduit. Make a hole through the ceiling or floor as necessary.

Having identified the appropriate circuit cable, cut it and insert a three-terminal junction box. Add the branch cable, connecting like core to like core. Alternatively, you can connect directly into the back of a power socket or to a loop-in ceiling rose.

Run the cable to the connector unit, connecting the cores to the appropriate terminals. Don't forget to run an earth core from the terminal of the mounting box to the earth terminal of the faceplate. You can make good the wall before connecting the flex from the shredder unit to the faceplate. Check that the correct size fuse is fitted, fit the faceplate and restore the power.

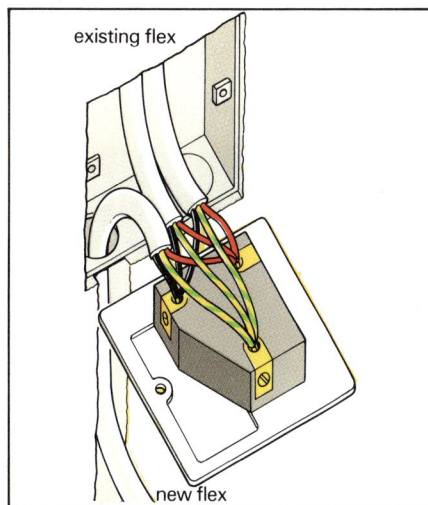

To wire into a ring circuit remove the face of a convenient socket. Check to see that there are two sets of three core cable and connect the matching colours together

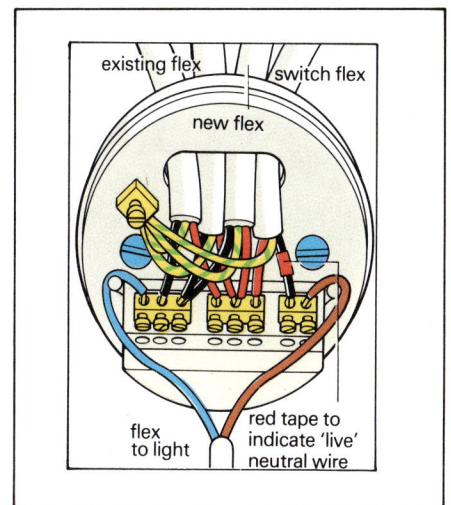

It is possible to wire into a loop-in ceiling rose by doubling up like cores in the supply terminals. However, it is easiest to wire into the last rose on a line and use its entry set of terminals

1 *Fix a fused connection unit to the wall and run cable to it*

2 *Tap into the power supply at a joint box, socket or rose*

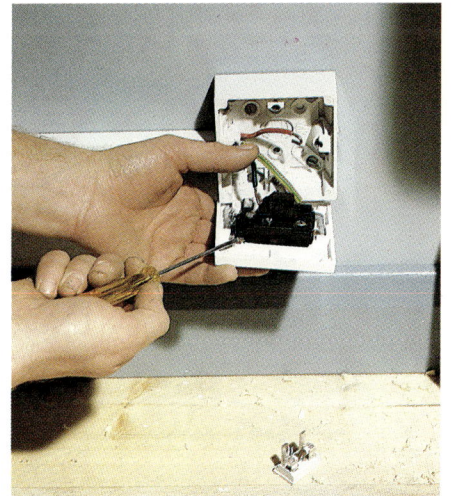

3 *Connect the flex to the supply at the connection unit*

CHAPTER 5

HOME MAINTENANCE

No matter how well your plumbing is installed and maintained, occasional problems are bound to crop up. But they can all be solved fairly easily, providing you locate the cause. A reliable set of tools is essential; make sure you keep them at the ready in case of emergencies.

PLUMBING TOOLS

Having the right tool for a plumbing job not only makes things easier, it also reduces the possibility of things going wrong. For example, it's all too easy to damage a nut with the wrong spanner, making it impossible to remove. Other jobs, like installing an immersion heater, just cannot be done if you don't have the proper tool. So knowing what's available and what you need for a particular job will really pay dividends.

Tools for jointing copper pipes

For compression fittings spanners are necessary; for capillary ones a blowlamp is needed. Compression fittings are easier to use and can be easily undone and reused if mistakes are made. However, capillary fittings are neater and the job of melting the solder to make the joint isn't difficult once you've mastered the technique.

For one or two joints only, a blowlamp with an integral gas cylinder can be used, but if you're making a large number of joints—as when installing central heating, perhaps—a **blowtorch** connected by a flexible hose to a large refillable gas cylinder will be much more economical. A flame-guard accessory for the burner is useful as it concentrates the flame around the joint and away from the surface behind. A glass fibre mat is a useful alternative for the job of protecting the surrounding surfaces.

To tighten the nuts on a compression fitting, two spanners are necessary—one to hold the fitting and the other to turn the nut. You're most likely to be making joints in 15mm and 22mm pipe, so it's best to have spanners in these two sizes. Add a couple of **adjustable spanners** to cope with any odd-sized fittings.

Adjustable pipe wrenches can be invaluable. It's worth owning a small pair; larger ones are expensive, and can be hired whenever you really need them. The ser-

Cut metal pipes with a saw or, more neatly, with a pipe cutter (centre). Bending pipes is easy with a pipe bender or a bending spring

rated jaws give a powerful grip, but may damage fittings if a lot of force is used.

Tools for cutting and bending pipes

For cutting pipes, both copper and plastic, you will need a **hacksaw**; a junior one will cope well unless you have a lot of cutting to do. Fit a fine-toothed blade. If you don't fancy hacking your way through a lot of pipe, buy a **pipe cutter**. This tool clamps the pipe between fixed jaws and a cutting wheel that trims the pipe neatly as the tool is rotated around it, leaves no swarf (rough metal) and can cope with all common pipe sizes—15mm, 22mm and 28mm.

If you're using a hacksaw, you'll need a fine **file** to remove the swarf and burr left after making the cut plus some steel wool to clean up the pipe ends before filling.

There are two tools for bending copper pipe. The first is a **bending spring**; this is a coiled steel 'snake' which slips inside the pipe to support its walls and prevent kinking while you bend it, and is then pulled out once the bend has been made. The springs come in different sizes to suit the commonest pipe diameters, up to a maximum of 22mm—pipes larger than this cannot be bent by hand.

For larger projects (and larger pipes) you need a **pipe-bending machine**, which you can easily hire. With this, the pipe is placed across the bottom plate of a former and is bent by the top plate as it is pulled down into position by a lever.

Using **unkinkable copper pipe** avoids the need for bending tools. This is pipe with straight ends for making connections, but with the centre portion made flexible by corrugations in the pipe wall. It can be bent without a spring or former and is useful for making tap connections where you want to bend the pipes in awkward situations—behind a basin pedestal, for instance. However, it is more expensive than buying ordinary pipe.

Tools for baths and basins

Taps are often tucked tight against a wall, making it impossible to use a spanner in the usual way. In these cases you will need a special **basin spanner**, sometimes called a **crowsfoot spanner**; it can be used hori-

Fit taps to baths and basins with a basin spanner (top), and immersion heaters with an immersion heater spanner (bottom)

zontally where space permits, but will also turn nuts when held vertically. There's a larger version called a **bath wrench** for bath waste back nuts, and a double-ended version that will cope with both sizes.

Another tool that can be used vertically is the plumber's **adjustable wrench** (not to be confused with a pipe wrench). This tool is far less common than the crowsfoot spanner and also more expensive, but it is very useful in certain situations. It is supplied with two or three universal heads (small—for 15mm nuts; medium—for 22mm nuts; and large—for 35mm and 42mm nuts) which can be used at more or less any angle to the shaft. It's a self-tightening tool which takes a stronger grip the more it is turned and will sometimes shift nuts that other spanners won't budge. As with other expensive tools, it's probably best to hire one when you need it rather than to buy it for your tool kit.

Tools for installing a WC and immersion heater

The tools you need for installing the bowl of the WC depend on where it will be sited and the material of the soil pipe to which the bowl will be connected. The WC-to-soil pipe connection is usually made with a flexible plastic connector which simply pushes on to the pipes and requires no special tools. The connection between the bowl and the cistern is usually a push-fit assembly too. You will need screws and a screwdriver for securing the bowl to the floor and fixing the cistern to the wall. You may need a **bolster chisel** and **club hammer** if you are removing an old WC which was mortared to the soil pipe.

To install an immersion heater you need all the usual tools plus a special large spanner to screw the immersion heater into the boss which is normally provided on the dome of a hot water cylinder. This spanner is 2¼in wide and may be open-ended or ring pattern. Builders' merchants usually sell them, or you can hire one from a local hire shop if you prefer.

General-purpose tools—including a power drill, screwdrivers, a file for smoothing off pipe edges and a handyman's knife—will be needed for most plumbing jobs

Tools for plastic pipework

Like copper, plastic is easily cut with a fine-toothed saw, but you're likely to be cutting much larger diameter pipes for soil and waste runs and it will be important to get the cut exactly square. A set square is useful for this, but you can get a very good line by wrapping a sheet of paper round the pipe and using its aligned edge to cut the pipe at right angles.

You'll need a handyman's knife to trim large pieces of swarf from the cut and a file to rub off the burrs. You may also need a small brush for applying the solvent-weld cement which glues the joint; however, many cements come with an applicator attached to the lid of the container.

You can also cut rigid cPVC (chlorinated polyvinyl cloride) pipe (used for water supply and central heating runs on some very modern installations) with an ordinary pipe cutter; for polyethylene and poly-butylene pipe, special shears are ideal but a hacksaw will do the job quite adequately.

To hire or buy?

If a tool is fairly cheap then it's usually worth buying rather than hiring. Although this adds a little to the cost of the job, you have the tool for future use and you can do the current job in your own time without worrying about hire charges clocking up.

Larger, more expensive tools are another story. It usually won't be worth buying a pipe bender, for instance, but might well be worth hiring one for a day or so if you've got a lot of work to do.

It is also well worth hiring the more specialised tools, like immersion heater spanners, which you're fairly certain you won't use more than once in a long while.

General-purpose tools and materials

Many of the tools in your toolbox will be useful for plumbing. You'll probably have a hacksaw suitable for cutting copper or plastic pipe (invest in a new blade if you've been using the saw for other jobs). A file is a must for smoothing pipe ends. For fitting pipe clips securely and accurately you'll also need a drill and screwdriver.

Plumbing can involve lifting floorboards to run pipes below the floor, so a saw will be useful and also a bolster chisel for levering

up the boards. A bolster chisel and club hammer will be useful if you have to chase walls, and if you're taking pipes through a wall you'll need a large masonry drill bit.

There are also plenty of sundries you will need. These are cheap to buy, so it is worth investing in supplies of the following:
• A roll of **PTFE tape**—a turn or two of this low-friction plastic tape will also help to give a good seal at joints.
• **Jointing compound**—a smear on the meeting faces of a fitting prevents leaks.

If you're using capillary fittings you will need **flux paste** and for plastic solvent-weld joints, you will need the special **solvent-weld cement**. **Hemp** is the only other specialist jointing material you are likely to need and then only for large joints and those near central heating boilers.

Pipe clips are essential with long runs of pipes; some come complete with screws or masonry pins. **Pipe lagging** in foam or glass fibre is useful for all hot water supply pipes and for cold pipes that rub through unheated areas—as well as insulating the pipe it stops condensation forming on cold pipe walls.

Plumbing emergencies

Burst pipes are the most common plumbing emergency and as they can cause a lot of damage, it makes sense to be prepared, by keeping an emergency repair kit.

A small emergency kit containing a sink plunger, some spare joints, tap washers and 'O' rings will prove invaluable

The first thing to do is to limit the amount of water that can flood out. Stop water coming into the house by turning off the main stopcock and drain the cold water cistern and pipes before tracing the burst.

Copper pipes often fail at joints and, if you're competent at joint making, your repair kit should include spare joints to fit in place of a damaged one. To remake a leaking compression joint you'll need some new olives in 15mm and 22mm sizes—the old one will be crushed, and will have to be cut off with a hacksaw. For a quicker but less permanent repair, which doesn't need tools, you can use a leak-sealing paste to patch up a joint. Most of these are two-part products—you mix together resin and hardener to get a working paste which will rapidly set to seal the burst. You usually use them in conjunction with some sort of re-inforcing tape.

You can make a more permanent repair with a burst pipe coupling—a length of pipe with a joint at either end—which you fit in place of the split length.

A plunger is a useful gadget to have in the house—use it for clearing blocked sinks and waste pipes. Lastly, keep a few washers to stop dripping taps, plus 'O' rings for mixer taps and solid washers for the ballvalve of the WC cistern.

LEAKS AND DRIPS

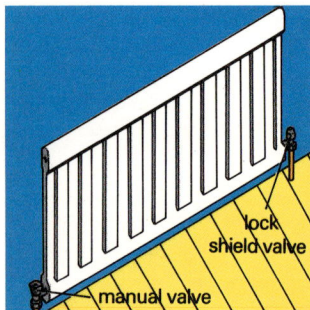

The radiator can be isolated for repair or replacement by turning off both valves

1 Tightening the nut under the manual (or wheel head) valve may be enough

2 If you don't have a lockshield key, use a spanner with care

3 Finally, turn off the manual valve at the other end by hand

Curing a leaking radiator

A leaking radiator should be looked at without delay—to avoid damage to carpets and floorboards. If it is leaking at the coupling to the pipe there's little cause to worry. Tightening the nut may be enough to stop the leak. If not, you need to undo the cap nut, take off the fitting, then replace the olive.

You'll have to drain the radiator before you can tackle this job. And if you need to carry out repairs to the valve, you'll have to drain at least part, if not all, of the system.

If the leak is caused by corrosion, the only cure is to fit a new radiator. You can make a temporary repair, however, using a plastic resin filler.

To work on the radiator you will need to isolate it, by turning off the valves at each end. To turn off the lockshield (see diagram), remove the cap which holds it in its set position then fit a key onto the top.

Next turn off the manual valve at the other end. You can then take the radiator out to replace it or carry out a temporary repair more easily.

If you install a new radiator, you can avoid further corrosion problems by adding a rust inhibitor to the water after flushing out the system with clean water.

Replacing a WC connection

A leaking joint between a lavatory pan and its soil pipe should be attended to immediately—even if you can only carry out a temporary repair.

As an emergency measure, clean and dry the area thoroughly and wrap it tightly with heavy duty sealing bandage such as Sylglas. This is a temporary solution, so remake the joint as soon as possible.

The soil pipe will be connected to the pan in one of three different ways.

The oldest method used for earthenware and iron soil pipes was to caulk the joint with tarred hemp and then fill it with cement and sand mix.

In later years these joints were often packed with putty; most recently plastic connectors have been used to join the pan and soil pipe.

To reseal a sand and cement joint can pose some severe problems. It entails chopping out the cement, which is very difficult to do without breaking either the pan outlet or the collar of the soil pipe.

If the joint is packed with putty this will have set hard but it is not too difficult to scrape out using an old screwdriver.

The easiest way to remake the joint is to fit a plastic Multikwik soil pipe connector. To do this, you'll need to take the lavatory pan right out.

Unless you are dealing with a close-coupled lavatory pan start by disconnecting the flush pipe at the back of the pan. This is usually held by a plastic sleeve which you

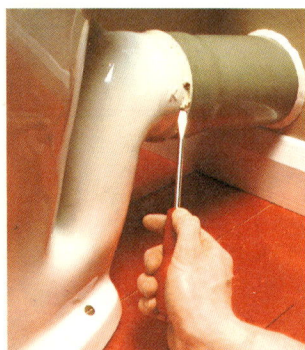

1 If putty seals the waste joint, scrape it off with a screwdriver

2 The pan may wriggle off the mortar base, or it may be screwed

3 Fit the Multikwik connector into the end of the soil pipe, then to the WC

4 Use the grease—it will make the job easier and help get a better join

simply roll back.

Remove the screws holding the base of the pan to the floor and lift the pan out of the way—be careful, though, it's heavy. If the lavatory has a solid floor, the pan will probably be bedded into a mound of mortar to keep it level. The only thing you can do in this case is to try to wriggle the pan free—prising it from underneath could break it.

Clean the outlet of the pan, and the collar of the soil pipe before fitting a flexible plastic connector such as a Multikwik. One end of this fits into the soil pipe socket. The other end fits over the pan waste outlet.

You can then replace the lavatory pan and screw back to the floor. The plastic connector has enough flexibility to allow for any slight discrepancies in the alignment of the pan outlet and the soil pipe.

Resealing leaking pipe joints

The weakest points in any pipe-work are the joints. If copper pipe is involved, resealing a leaking joint is either a simple matter of tightening a nut, or soldering a joint. Joints in lead pipe are not really a DIY job.

Copper pipework is often connected by compression joints. With these a slight turn of the nut will generally compress the olive inside sufficiently to cure the leak.

Soldered joints are more of a problem as they cannot be resealed without first emptying the pipe and this means draining at least a part, if not all, of the water system. If you are dealing with a hot water pipe (other than central heating), tie up the ball valve in the cold supply tank so that the water can be drained off at the taps. For central heating turn off the boiler and open the drain cock. A cold water pipe only needs the water turned off at the main stop tap. But even with the system drained, there may still be some water in the pipe.

There's no point in simply trying to re-solder the joint—the pipe will be dirty inside the fitting, and you need a clean and shiny pipe if the solder is to grip. You also need a new capillary joint that does the same job as the old. But you may not need an identical fitting: if the original was the type that uses a separate solder supply, buy the pre-soldered kind instead as they are very much easier to use.

The best method of taking the joint apart is to make a cut right through the centre of the fitting with a hacksaw and let any water out. Then heat the pipe fitting with a blow-torch to melt the solder; this will allow you to pull out the ends of both sections of pipe. You may have to loosen some pipe clips in order to spring the pipe enough to allow you to free it.

When using the blowtorch, you must take great care not to damage the wall behind the pipes. The best way to do this is to tape a piece of asbestos to the wall. The type of asbestos mat used by plumbers is expensive, and an excellent substitute is an asbestos simmering pad found in most good kitchenware shops.

Clean the pipe thoroughly with wire wool or abrasive paper making sure that all the old solder is removed. Smooth the cut ends with a file.

Coat the pipe with flux to keep it clean. Clean the new fitting before sliding it over the ends of the section of pipe.

When it is in place heat it with a blow-torch until the solder appears around the

1 *Start by cutting straight through the centre of the joint at right angles*

2 *The old joint will come away once the solder is melted with a blowlamp*

3 *Clean the pipe thoroughly with wire wool or glass paper to remove grease*

4 *Then coat both ends of the pipe with flux. Apply flux with a matchstick*

5 *Protect the wall behind when using a blowtorch with a special mat*

6 *First push the nut, then the olive, onto the pipe. Apply jointing compound*

7 *Next fit both ends of the pipe into the fitting. Push them hard into the ends*

8 *Finally, tighten up by hand then use an adjustable spanner*

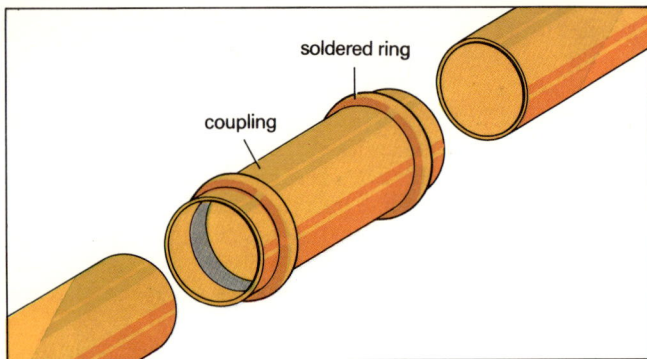

Presoldered joints are easy to fit, but must be heated

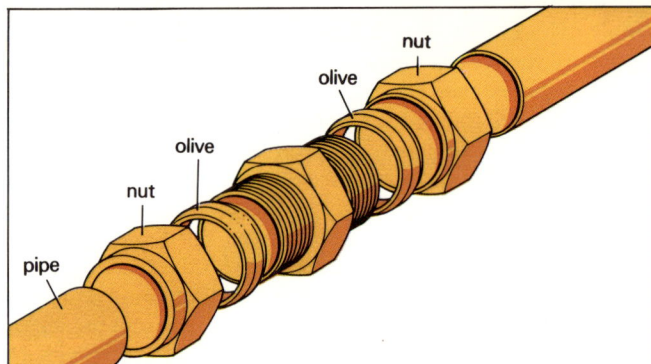

Compression joints can be undone with spanners

end of the fitting, indicating a fully soldered joint. Remove the heat as soon as you see the solder appear. Never continue to heat the pipe after this point.

An easier solution may be to replace the existing solder joint with a compression fitting which needs no heat.

The compression fitting has a nut at each end with a ring, or olive, inside. First push the nut and then the olive onto the end of the pipe which must be clean and smooth.

Push the pipe into the fitting as far as the moulded stop, then push the olive into the end of the fitting.

As you screw the nut up, the olive is forced into the fitting and compressed onto the pipe, forming a water tight joint. Tighten the nut as far as you can go by hand—then give it a further 1½ turns with an adjustable spanner.

Lead pipe is more difficult to repair. You must scrape the pipe clean and cover the bright area with tallow flux. You then melt a plumber's solder onto the pipe and wipe it in to the traditional shape using a coarse cloth coated with tallow.

Don't tackle this job unless you feel confident you can handle it—use an epoxy filler to make a temporary repair, then call in an experienced plumber.

Dealing with a leaking overflow

Although you only fill a sink up to the overflow infrequently, it's important that overflow connections are well sealed. A leaking connection will cause a great deal of mess but it's often difficult to determine whether it's the overflow connection or the sink waste that's leaking. So before you do anything, fill the sink up to the overflow outlet and check underneath to see just where the water is coming from.

Most modern kitchens and inset sinks have a flexible pipe leading from the overflow outlet to the waste pipe, just above the trap connection to the sink. If the leak is at the connection to the outlet in the side of the sink, take off the pipe by unscrewing the little grid inside the sink.

On some of the more recently produced kitchen sinks, the grid is incorporated in the sink itself and to undo the overflow pipe to reseal the connection, all you need to do is to unscrew the nut which the plug chain is attached to. And a few sinks don't even have an overflow outlet.

Clean the outlet and the end of the pipe then screw the grid back in position, making sure it's as tight as possible. If the leak is around the grid, seal it with a little silicone sealant before screwing it back.

When you cut the nozzle of the sealant tube, make sure the opening is not too wide. For once you apply the sealant and tighten up the grid, if you have applied too much, it will ooze out of the side of the grid and may look unsightly.

The other point where a leak can occur is at the connection to the sink waste. This connection may push on or take the form of a collar which fits around the metal outlet from the sink. To undo the connecting nuts, you will need a large adjustable spanner or wrench.

Undo the connection and clean the joint thoroughly. Check the order in which any washers are fitted as you remove them and also check and replace any damaged ones. You can then reconnect the waste and overflow and test the connection with a sink full of water.

These joints are not under any pressure, so they should be water-tight without excessive tightening of the nuts. If the connections still leak once the joints have been cleaned and re-tightened, undo the defective joints and apply a thin squeeze of sealant before reassembling the connections.

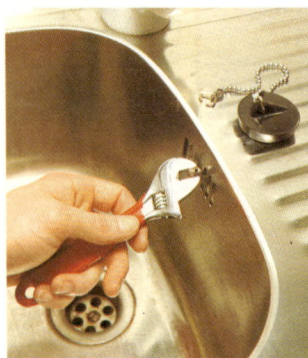

1 *On some sinks, there's only a screw holding the overflow. Undo it with pliers*

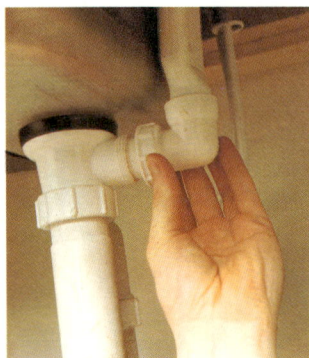

2 *Loosen the waste fittings with an adjustable spanner, then free by hand*

3 *Don't cut the nozzle too large to avoid too much sealant*

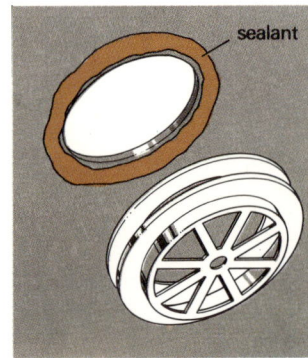

On modern sinks you buy the whole overflow fitting as one unit

WC AND BATHROOM REPAIRS

Clearing a blocked trap

When a bath or basin becomes blocked it is usually the trap underneath that is at fault. Loose hairs are the most common cause of trap blockages and, if conventional plunging fails to shift them, you have no choice but to remove the offending material by hand. How you do this depends on the type of trap installed. If you are dealing with a bath, you will have to unscrew and remove the panelling to gain access to the trap.

Old-style U trap: These are generally of brass or lead, and must be handled with care to avoid damage.

Start by placing a bowl underneath the trap and a piece of wood in the U part. Holding the wood in one hand, to counteract the turning force, use a wrench to unscrew the clearing eye which you will find at the base of the trap.

Hook out any debris remaining in the trap with stiff wire. Then clean the thread of the clearing eye and wrap a few turns of PTFE joint tape around it. Replace the eye, turning it a fraction over hand tight.

Plastic U trap: In this case you must dismantle the entire trap.

Unscrew the locknuts either side of the trap by hand, wrapped around a cloth. If they won't budge, boiling water should shift them.

When you reassemble the trap, a few turns of PTFE tape around the locknut threads will prevent leaks.

Bottle traps: These are the simplest of all to clear.

Hold the waste pipe in one hand and unscrew the cover with the other, wrapped in a piece of cloth.

Having allowed the debris to fall out into your waiting bowl, poke some stiff wire into the waste pipe to clear any residue then refit the cover. Again, a little PTFE tape wrapped around the cover thread will prevent leaks and save you having to tighten the cover too hard. Run the taps to check the repair. Also check that the waste is tightly connected and does not leak.

Changing a tap washer

When a tap starts to drip from the spout, it's usually the washer that's at fault. Replacement washers are available cheaply from hardware stores but, like the taps they fit, most are still in Imperial sizes—½in for basins, ¾in for baths. Only imported Continental models and the more unusual modern designs are Metric, the equivalent sizes being 15mm for basins and 22mm for baths and sinks.

Tap washers seldom fail when the shops are open so it pays to dismantle one of each of your tap types while they are still in good order and take the washers to your local stockist so that spares can be matched up.

Start the job by isolating the water supply to the tap concerned. Look for a stop valve on the pipe supplying the tap and turn this clockwise to cut off the water. If there are no such valves, what you do to drain the system then depends on whether the tap is for the hot or cold water supply.

Cold taps: If your plumbing system includes a cold storage cistern, look for a stop valve on the supply pipe running from the base. Otherwise tie up the ball valve to stop water entering the cistern, then open all your cold taps to drain it. (Note: this

applies to bathroom cold taps; kitchen cold taps are plumbed direct to the rising main, in which case turn off the main stop valve to isolate them.) If you have a direct system with no cistern, simply shut off the main stop valve and open all the cold taps. With this type of system you will only wait a few seconds for the pipes to drain.

Hot taps: Look for a stop valve on the cold supply entering the base of your hot water cylinder or water heater. You may be surprised to find that this stops any hot water leaving the cylinder: simply open the tap being repaired, to drain any water still left in the supply pipe.

To get at the washer, start by removing the tap handle. Old-style crutch-type handles are held on by a small screw—either under the colour button, or underneath the handle where it joins the tap stem. Follow by unscrewing the shroud below the handle with a pair of self-grip pliers—put some cloth in the jaws of the pliers to protect the tap chrome.

1. New-style plastic handles either simply pull off or are held on by a screw fitted under the colour button.

2. The next step is to unscrew the entire stem assembly from the tap body. Get a spanner, self-grip pliers or a plumber's wrench around the large nut where the two meet and apply force. As you do so, take care not to move the tap body itself or you

★ WATCH POINT ★

If you do not have a replacement washer make a temporary repair by turning the old one round.

1 *Remove handle*

2 *Unscrew stem assembly*

3 *Release the old washer*

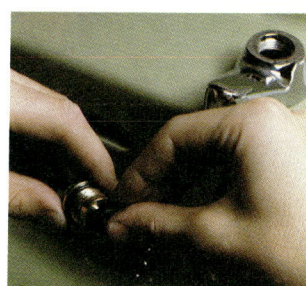

4 *Fit new washer in its place*

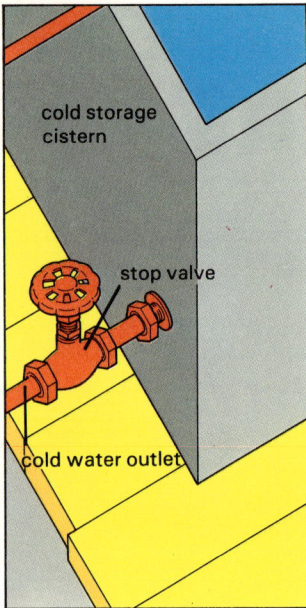

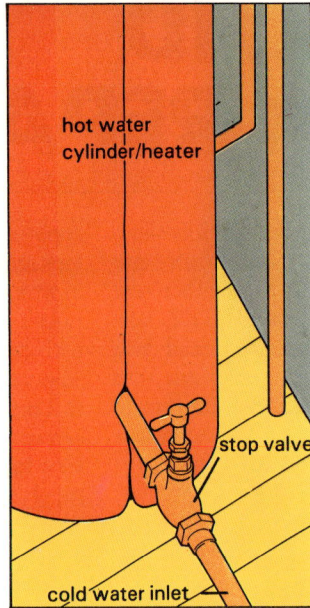

Stop valves: cold water cistern (left) and hot water cylinder

1 *Remove pin to release arm* **2** *Unscrew piston cover*

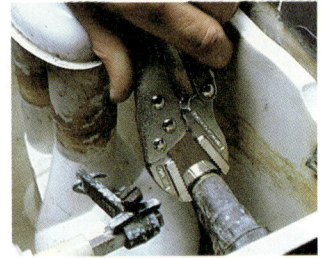

3 *Juggle out and remove piston* **4** *Dig washer out of piston*

may disturb the pipe joint below. If the nut is tight, wedge the body with a length of wood against something solid so that you counteract the turning force.

★ WATCH POINT ★

If your tap continues to drip, the washer seating is worn. The conventional solutions are to replace the tap or regrind the seat using a special tool. The easy way is to fit a 'HoldTite' domed washer or a push-in nylon seat—ask for them at your plumbing supplier.

3, 4. You will find the offending washer on the end of the stem unit. The older rubber sort are held in a mounting plate by a small nut or screw. Undo this and dig out the washer, then slot in the new one, replace the nut and tighten.

Newer nylon washers simply snap into place on the end of the stem or onto a mounting plate—prise the old one off with the end of a screwdriver.

In both cases, reassembly is an exact reversal of the dismantling procedure. Turn on the water and test.

Toilet cistern overflow

Overflows here are caused by a failure of the ball valve washer controlling the water flow.

Most probably your problem valve is an old one. In this case the replacement washer will be rubber, in an Imperial size—almost certainly ½in (to match the supply pipe).

Start the repair by shutting off the water supply to the ball valve.

1. Now disconnect the ball arm from the valve by removing the split pin and then juggling the end free of the washer piston.

2. With the arm removed, unscrew the piston cover.

3. Stick a nail in the slot where the ball arm entered the valve and flick the washer piston out through the end of the valve.

4. On the other end of the piston is the washer. Dig it out of its seat and fit a new washer.

Fitting a new toilet seat

You can rarely repair a damaged toilet seat satisfactorily—it is usually easier to replace

1 *Slide end block onto bolt and tighten from below*

2 *Fit pivot rod into block and fit on other block*

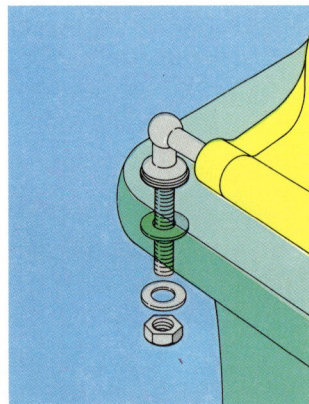

Ball-headed bolts are generally found on older fittings

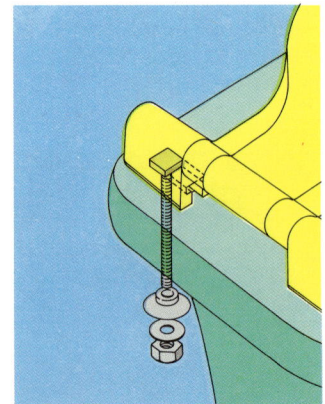

Modern fittings are usually the sliding end block type

it. Seats and lids are normally sold as a set, though some manufacturers will supply them separately. Remember that some lids are only designed as covers and are easily damaged if any weight is put on them—others are sturdier, and will withstand the weight of the average man.

1. To fit a new seat, you need to bolt it through the existing holes in the back of the pan rim (see diagram). Check the diameter of the bolts and the distance between them (if this is not adjustable) to make sure the new seat will fit your pan.

2. If the damaged seat is bolted into place with metal bolts which are rusted over, you will need to soak them in penetrating oil before trying to remove them. Modern units are fitted with plastic bolts which will not rust and should always be tightened only finger tight.

A common problem in many lavatories is that the seat fails to stay up as the cistern sticks out too far. If you have this problem, ask whether you can get a cranked fixing bolt for the new seat. This allows the seat to be moved forwards by up to 50mm, while using the existing holes in the pan rim to bolt the fittings securely in position. Fitting is simple and quick.

Unblocking a WC

The correct tool for dealing with this is a cooper's plunger—a plain rubber disc on the end of a T bar—NOT a plunger with metal disc, which would damage the pan.

You can get by using a conventional sink plunger, though this will not be the right shape to create full pressure—it can only force short, sharp surges of water through the trap to clear the blockage.

To this end, insert the plunger in the trap and move it up and down vigorously in 30

household mop is a good substitute. Or you could try wrapping a toilet brush in rags and then enclosing it in a plastic bag tied with string. Use both as you would a plunger and see if this frees the blockage. If not disconnect the trap and clear the U bend by hand.

Flush pipe repairs

The flush pipe that connects a WC cistern to the pan is liable to leak from either of its connections at some time.

At the cistern end the water will seep through the nut and run down the pipe. Often the first sign of a leak is a wet patch behind the WC pan which may be mistaken for a leaking soil connection.

If you're not sure where the water is coming from, wrap a piece of toilet tissue around the flush pipe at the top end and then flush the cistern. If the tissue is wet you may safely assume the source of the leak is above it.

There is no absolute need to empty the cistern to carry out repairs to the flush pipe and, therefore, no need to turn the water supply off or tie up the ball valve provided you do not disturb the backnut on the cistern siphon.

Gently try to undo the lower nut securing the flush pipe on the threaded section protruding through the bottom of the cistern. If it's an old nut that has been painted, this may be difficult. Rather than exert undue force, apply paint stripper around the nut and leave it for a few minutes. In most cases, the nut will then undo easily.

Once you have slipped the nut down the flush pipe you can examine the seal. Old cisterns may have hemp wrapped around a swivel joint which looks like a large tap connector.

(available from plumbers' merchants) several times around the connection. It's a good idea to knot it in position. Smear some plumber's putty around the hemp and re-tighten the nut.

On some copper and plastic pipes, you'll find a rubber O-ring instead of a hemp joint. Between the rubber and the backnut is a ring of plastic or metal which prevents the rubber from being distorted when the nut is tightened.

If the rubber seal is perished or damaged,

replace it with a new one. Before tightening the backnut, smear a little silicone grease or washing-up liquid around the assembly. Plastic nuts can usually be sealed by hand—a piece of rag wrapped around the nut will help you gain a good grip.

If you have to use a wrench on a plastic fitting, do so with care as you may easily distort the nut. There is no pressure in a flush pipe, so the seal doesn't have to be too tight. If the leak is occurring at the lower end of the flush pipe, you'll probably find that a rubber joint called a flush cone has perished. To remove the flush cone you will have to gain some free movement on the flush pipe by undoing the top connection.

Once the flush pipe is loose, gently work it out from the back of the pan and remove the perished rubber cone.

Universal rubber cones are available from plumbers' merchants. If the one you buy

Using a plunger to siphon the WC trap

1 *The seal round the flush nut may need replacing*

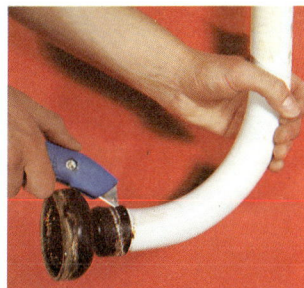

2 *Use a knife to remove a perished flush cone*

3 *A new cone can be made to fit by inverting it*

second bursts. When the blockage clears, flush the cistern to refill the trap.

If you haven't got a plunger, an ordinary

Carefully cut away all the old jointing material and leave the mating surfaces clean. Then wrap a new length of hemp

doesn't fit your pan, simply turn it inside out to give you another size. Slip the flush cone onto the lower end of the pipe first and insert the pipe into the pan. You can then pull the cone forward, easing it onto the chinaware as you go. A lubricant such as washing-up liquid will help to make this procedure a good deal easier if the cone is a fairly tight fit.

Some old installations with lead flush pipes have a putty and bandage connection at the pan end. Often the china socket is not evenly formed and a new rubber cone fails to make a good watertight seal.

In this case, cut away all the old putty, being careful not to damage the pipe. Repack the joint with linseed oil putty and wrap a new gauze bandage around the joint in a similar fashion to the old one. Truss the joint with string to hold it in place while it dries. After a week, apply an oil-based paint around the whole joint to make a final seal that is also easy to wipe clean.

If you find the old flush pipe is faulty you can replace it with a new plastic assembly which will need no further maintenance once installed.

High level flush pipes come in two patterns to suit rear-mounted or side-hung cisterns—the pieces slot inside each other and need no sealing compound. You may have to shorten the spigot ends of the pipe to suit your installation. This is easily done by cutting with a fine-toothed saw.

Renewing a cistern

If your cistern is cracked or rusted through it will have to be replaced.

To make the job easier try to select one that matches the dimensions of the old one. The important measurement is that between the overflow pipe and the top of the flush pipe, since it will save you having to make alterations to the plumbing.

Turn off the water supply and drain the old cistern.

Disconnect the cold supply and oveflow pipe and undo the flush pipe connection.

Cast iron high-level cisterns are quite heavy. To avoid accidents, hand the cistern down to an assistant on the floor.

New cisterns come with their workings packed inside, so before connecting to the plumbing the cistern must be made up. This entails fitting the siphon and flush handle assembly. To determine whether this goes on the left or right you will need to fit the ball valve to match your existing supply. The ball valve and overflow connector are interchangeable in their respective holes. Once these are in position, it will become clear which way the siphon fits.

On most new cisterns a dual flushing option is available to comply with local water regulations. This allows only half the water content of the cistern to be discharged by releasing the handle immediately after flushing. To obtain a full flush the handle must be held down until flushing ceases. If you do not require a dual flush facility a small plug must be fitted in the top of the siphon.

When the cistern is fully assembled, offer it up to the pipework and mark the position of the brackets and fixing holes in the back.

Drill and plug the holes, making sure they are level. The cistern can then be hung in position on the wall.

Screws inside the cistern must be rustproof—either electroplated or brass.

When you are satisfied that the cistern is secure connect the supply to the ball valve and the overflow connector.

On plastic ball valves, wrap a few turns of PTFE tape around the end of the threads as well as fitting the fibre washer.

Connect the flush pipe, taking care not to twist the siphon out of line with the flush lever. Finally, turn on the supply and adjust the water level to coincide with the mark in the cistern.

> ### ★ WATCH POINT ★
>
> On china cisterns use a couple of tap washers to protect the cistern from being damaged by screw heads.

Improving flushing

Poor or unreliable flushing can be attributed to a number of different things.

First make sure the water level comes up to the line indicated on the inside of the cistern. If it is too low, it can be raised by adjusting the ball valve.

Adjustment on brass valves can be achieved by bending the arm upwards. Hold the arm near the valve body with one hand and bend up gently with the other.

The flushing mechanism is fairly simple. When the handle is pulled down, the plunger pushes water through the siphon and into the flush pipe. As the float drops with the water level, the ball valve opens and draws in water from the supply pipe

1

2

3

1 *Fit a plastic plug over the siphon to prevent dual flushing*

2 *Support a low level cistern on your lap while offering it up*

3 *Wrap some PTFE tape round plastic ballvalve shanks*

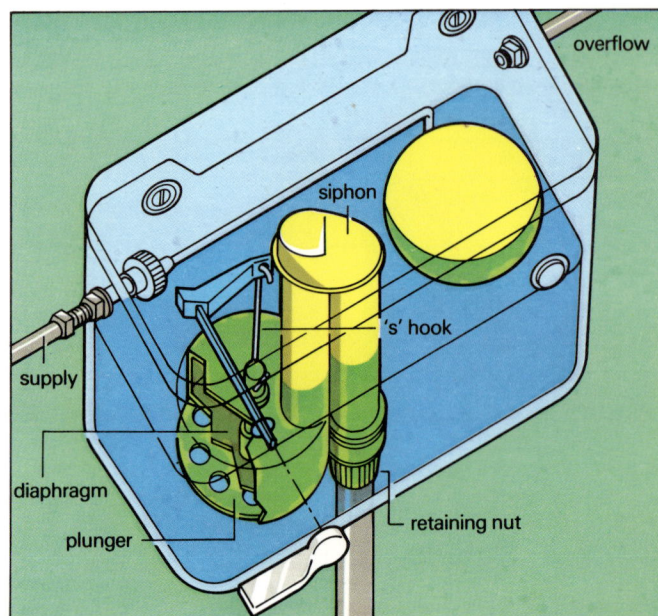

overflow

siphon

's' hook

supply

diaphragm

plunger

retaining nut

Plastic ball valves have an adjustment screw either at the float end or near the valve body.

Once you're satisfied the water level is correct, try the cistern again. If it still doesn't flush satisfactorily, the fault is probably with the diaphragm inside the siphon mechanism.

To change this you will have to empty the cistern. Turn off the supply to the ball valve and remove the water from the cistern. If you cannot flush it, you may be able to siphon it with a short hose pipe which you can discharge into the WC pan. Alternatively, bale out the cistern with a small pan and mop out any remaining water with a sponge.

Once the cistern is completely empty, undo both nuts underneath—this should leave the siphon mechanism free to pull up and out. You will also need to detach the siphon mechanism from the lever mechanism. Usually, it is possible to manoeuvre the brass 'S' hook out of the centre spindle—the siphon should then be easily removable.

Examine the siphon for cracks or holes. If it appears faulty, take it to a plumber's merchant and purchase a replacement of the same size.

If the siphon is sound, the likelihood is that the rubber or polyethylene diaphragm is torn or perished.

This is removed by sliding the plunger backplate and spindle out of the housing.

If the diaphragm is in any way misshapen or torn, you'll need to replace it. A new diaphragm can be purchased from a plumber's merchant, but should you have any difficulty obtaining the correct size and shape you may find it easier to make one.

A heavy gauge piece of polyethylene from something like a fertilizer sack will make a perfect diaphragm. Simply cut out a new one with some scissors using the old one as a template.

When re-assembling the diaphragm, remember to replace the small rubber washer that holds it down on the spindle. Check the seal on the underside of the siphon. If it's perished, it's well worth replacing while you have the cistern apart—ensure that no dirt is trapped under the seal as it might cause a leak.

On porcelain cisterns, a protective washer must be fitted between a metal back nut and the cistern to prevent cracking.

Close-coupled WC cisterns sit directly on the pan without the need for a flush pipe.

Occasionally, a leak may occur where the cistern meets the pan. First check that the two wing nuts underneath the pan are secure. If the leak persists, you will need to replace the rubber seal.

Shut off the water and drain the cistern. Disconnect the overflow pipe and remove the screws holding the cistern on the wall. Next undo the wing nuts and lift the cistern off the pan.

After purchasing the correct replacement, fit the new washer around the back nut so the threaded section of the siphon is poking through.

Ball valve troubles

On modern WC cisterns, the margin between a correct water level for satisfactory operation and a leaking overflow is very little. It is therefore necessary to maintain the ball valve in perfect working order.

If your cistern is overflowing, turn off the water supply and remove the ball valve assembly—but first check the float itself has no water in it.

Older style ball valves made of brass may have hard water scale deposits stopping their smooth operation. Dismantle the ball valve by removing the split pin and sliding out the piston. Gently remove all traces of debris from the moving parts of the ball valve. You can do this with fine emery paper or a proprietary scale remover such as the type used on kettles.

Check the rubber washer in the piston—if it is even slightly perished, replace it with a new one. The piston unscrews into two parts, enabling the washer to be removed, but these may be somewhat corroded. A good pair of adjustable grips will help you do this, but take care not to squeeze the piston out of shape.

Next check the seating onto which the washer closes. If this is pitted or worn, even a new washer will not seal it. Often the water will have worn minute grooves into its edges. In most cases the seating is replaceable with a new hard plastic insert.

New inserts come in three sizes. The smallest hole is for high-pressure cisterns fed directly from the mains, a middle-sized hole is for low-pressure cisterns fed from the cold water tank in the roof space, and the largest hole is for 'fullway' ball valves installed in flats where the cold water storage tank is only a couple of feet above the WC cistern.

If you use the wrong size orifice, you may experience problems with the cistern filling too slowly—which is annoying—or too quickly, which can result in a perpetual flushing of the pan. This happens when the cistern is filling so fast that the siphon never

has a chance to draw air because the water level is always too high.

Newer style plastic ball valves are now common on many WCs. Some designs have proved unreliable and have since been discontinued. An indication of this is the difficulty experienced when obtaining a replacement washer. Most WC manufacturers now fit plastic valves of a good design. A small screw is located at the end of the float arm to give precise adjustment of the water level in the cistern.

Plastic ball valves generally have the advantage of being quieter than the older style brass variety. This has obvious advantages for WCs.

If you wish to replace a brass valve with a plastic valve there are a few details which you should consider.

The new ball valve arm must go around the siphon—plastic arms are rigid so they cannot be bent to achieve this. A Torbeck valve has a very short arm only 50mm or so long—this is often a way of overcoming the space difficulty.

The threaded section of the ball valve known as the tail comes in 38mm and 50mm lengths; the longer tails are necessary on china cisterns, which have

★ WATCH POINT ★

When fitting a new ball valve, incorporate a small Ballofix-type stop-valve into the supply pipe. This will give you a means of isolating the ball valve during future maintenance.

particularly thick walls.

Ball valves are either bottom or side entry. Bottom entry valves are attached to a stand-pipe. The cistern will have to be drained to change this. Often you will find the new valve doesn't reach the existing pipe-work, but a 12mm tap extender may overcome this problem.

Oil-based sealing compounds should not be used on fittings with plastic threads.

It's a simple matter to fit one of these valves into an existing pipe run. Measure the depth of the shoulders within the fitting, subtract them from its length, cut out this much from the pipe and insert the valve. Tighten the couplings.

Dirt in the ball valve: small pieces of rust and scale tend to collect in ball valves, causing slow filling and damage to the washer. A small filter may be fitted in the pipework to collect the debris without interrupting the supply.

SILENCING NOISY PLUMBING

Air in pipes

Vented systems such as central heating and the hot water supply can suffer from air problems if they are not well designed. If your system suffers from this, the symptoms will be all too apparent.

In hot water services you may notice spluttering taps especially when running large quantities quickly. Usually, this is due to air being drawn down the vent pipe. This happens when cold water from the tank is not flowing fast enough into the bottom of the hot water cylinder.

Check first that the valve controlling this supply is a gate valve, and not a stopcock, which makes it unsuitable for low pressure systems. If you have the right type of valve, ensure it is fully open. Too many bends in the pipe run will also result in a poor flow.

A quick test to check whether air is being drawn down the vent pipe is to run the hot bath tap and at the same time gently place the palm of your hand over the open end of the vent pipe in the loft. If you detect any sucking this is conclusive proof; take your hand away immediately before any damage occurs to the plumbing system.

Cylinders are designed to take a 28mm cold-feed pipe. If yours is smaller, this may be the problem, and you may need to increase the flow by replacing the pipe with a larger one.

A build up of scale in the cylinder may also contribute to a poor supply. There are no easy answers to this problem. The entire system will need to be drained and the cylinder disconnected. It can then be flushed through with a hose pipe.

Air in central heating systems

Air in central heating systems is a major problem. It causes banging noises, howling or even whistling in cast iron boilers, and also increased corrosion of the boiler and radiators. If you find yourself having to bleed air from radiators at frequent intervals it is an indication that something is wrong with the system.

The best way to overcome this problem is to fit a de-areator to the vent pipe. If you do not wish to use soldered capillary joints choose a de-aerator which is suitable for connection to compression fittings. The unit itself is made of copper and is shaped rather like a bottle.

It works by causing turbulence which 'beats' out the air in the water. It is fully automatic in operation and requires no maintenance. The additive should be mixed according to the maker's instructions and added to the expansion (header) tank for the central heating.

Anti-corrosion additives for central heating systems will help to quieten down boiler noises caused by air on rough castings. They will also preserve steel-panelled radiators.

1 *Drain the system before you cut into the vent pipe. Allow plenty of room*

2 *Fit the de-aerator according to the maker's instructions. Make the joints strong*

3 *Bleed air from pumps on a horizontal pipe run by turning the bleed screw half a turn*

A typical central heating system showing the general layout, pipe sizes and the direction of flow of the water in the pipes. The boiler warms the cylinder, too

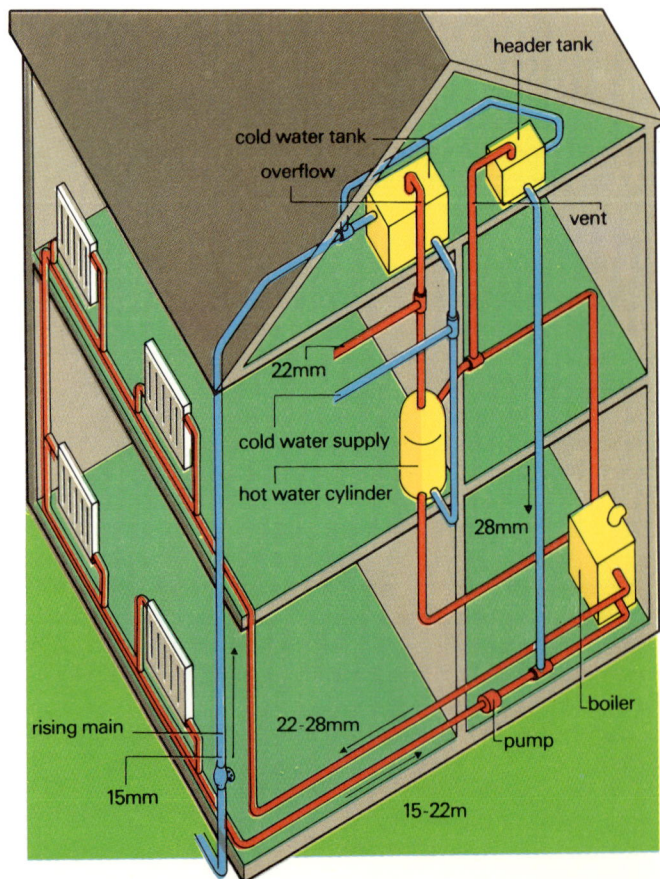

header tank

cold water tank

overflow

vent

22mm

cold water supply

hot water cylinder

28mm

rising main

22-28mm

boiler

pump

15mm

15-22m

Eliminating water hammer

When a fast flow of water through a pipe is halted abruptly (by turning a tap from full on to full off quickly, for example), shock waves are set up which carry back along the pipe and sometimes result in an audible banging. This *water hammer*, as it's known, often becomes progressively louder as the shock waves bounce back and forth and only quietens down very slowly.

Common causes of water hammer are faulty taps or stop-cocks, too fast a flow of water, and a faulty ball valve in the cold water storage tank. The problem can be aggravated by loose pipe securing clips or pipes which are not secured at regular enough intervals.

Curing water hammer is not always drawn into a string, around the spindle pushing it down in to the body of the tap. Then screw down the gland nut, tightening until the spindle is firm. The idea of this is to stop the spindle turning too quickly causing the flow to stop suddenly.

The jumper is the brass or plastic plate holding the washer in place. Take the tap apart as if you were changing a washer and remove the jumper. It should slide in and out of the spindle but have no appreciable sideways movement. Replacement tap jumpers are available from plumbers' merchants. Alternatively, you could fit a new tap and body.

Stopcocks suffer from the same problems as taps. The packing becomes loose and jumpers wear out.

Servicing a stopcock is no more difficult than a tap but you will have to shut off the but the pipes will be weakened.

Manufacturers overcome this by fitting flow restrictors inside the connecting pipes. The installation instructions supplied with the machine will indicate their whereabouts and may also tell you how long the machine should take to fill. Reduce the flow at the valve where the flexible hose joins the main plumbing if the machine is filling too fast.

Similarly, electric showers should have some restriction upon the input. Many manufacturers recommend the fitting of a valve which does not have a jumper. The Ballofix valve is one of these, and may be fitted at a convenient point in the supply. If the pipework is surface-run chrome-plated or stainless steel tube you can buy a chrome-plated version of the Ballofix.

Faulty ballvalves: Where water hammer only occurs when a ballvalve is operating,

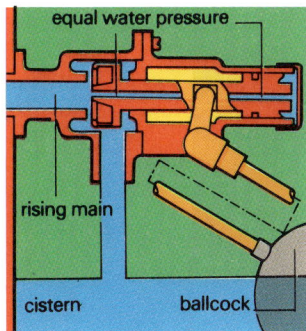

The equilibrium ballvalve uses water pressure in the valve body to eliminate the shock waves

1 *Tighten loose spindles on taps and stopcocks by tightening the gland nut with a spanner*

2 *If the jumper looks to be badly worn, replace it and the washer with new ones*

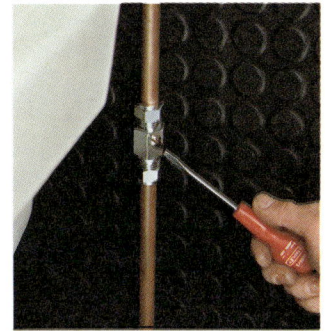

3 *Ballofix valves are very easy to fit, maintenance free, and easily adjustable by screwdriver*

straightforward, so you must be prepared for a little trial and error.

First, turn the stopcock down to reduce the flow. If this solves the problem and still leaves you with enough water at the taps, you need do no more.

More persistent water hammer will need further investigation. Start by fixing any loose pipes securely with clips.

Where pipes pass under wood floors, notches or holes are usually made in the joists to take the pipes. Use soft packing material—rubber, carpet pieces, rag, felt—to pad the pipes and help stop any vibration.

Taps and stopcocks: Noises heard when taps are running can be attributed to one of two things: loose spindles and worn jumpers. Before starting work turn off the supply to the tap concerned.

Tighten a loose spindle by turning the gland nut. If the spindle is still loose, undo the gland nut and slide it up the spindle. Wind a few turns of wool, or PFTE tape supply from outside the house. This is usually done at the Water Authority's stopcock just inside your boundary. Strictly, you should ask their permission to use it, but it is highly unlikely that they would object to you using it in an emergency.

If any one tap seems to be responsible for water hammer but there is no convenient way of restricting the flow to it, fit an additional stopcock.

If the supply pipe is on the surface of a wall, you can fit a special Ballofix valve rather than the more conventional stopcock. Ballofix valves are much neater in appearance and require no maintenance.

Solenoid valves such as those used inside automatic washing machines and instantaneous electric showers are designed to shut off in a fraction of a second. On mains supplies they can produce extremely violent water hammer.

With washing machines, the flexible hoses take most of the impact and you may not know that the hammer is taking place, the cause will probably be the ballvalve itself.

If a ball bounces up and down on the water's surface it will open and close the valve rapidly causing even more waves and more bounce. The result may well be extremely violent banging in the pipes.

Fit a damper to the ball to eliminate the bounce. Dampers are available from some plumber's merchants but you can make your own.

Wrap a stiff wire around the arm of the ballvalve just next to the ball. From this suspend a plastic disc or empty plastic yoghurt carton open end up. This will form a damper under the surface of the water and smooth its operation.

An equilibrium ballvalve, which is designed to operate under varying pressures, will also eliminate bounce without the need for a damper. These valves operate in the same way as a lock gate. By equalising the water pressure on both sides the valve will open and close with maximum smoothness.

A new version of this equilibrium valve is the 'Torbeck' valve which is silent in operation. It is therefore ideal for fitting to noisy-filling WC cisterns.

Modern PVC cold water tanks are less rigid than their steel counterparts. For this reason an alloy or galvanized steel plate is supplied to clamp the ballvalve to the tank. This provides increased rigidity around the mounting, so the ballvalve will be braced as it shuts off. The supply pipe to the valve must be clipped securely to a joist to eliminate any movement.

Similarly, many PVC tanks have substantial stiffening ribs on the bottom which prevent the water 'swilling about', and thus prevent ballvalve problems.

Severe water hammer

Severe cases of water hammer that resist cure by more simple measures can be dealt with by installing a form of shock absorber in the mains system. The ideal position for this is at the top of the rising main.

your kitchen tap and then refilling.

Similar commercial units are called Hydropneumatic accumulators. Unlike the home-made air chamber they contain sealed vessels of helium.

Blowing out airlocks

On indirect systems—that is those fed from a cold water storage cistern—occasional airlocks can be blown out by connecting a hose pipe between the affected supply and the cold water tap in the kitchen.

Turn on the air locked supply first, then the mains tap. It may take several minutes to blow out the air during which time a lot of 'gurgling' noises may be audible.

Air will collect at any high point in pipe runs. You must therefore make the pipes rise progressively to an open end. This means in effect that pipes should fall away from the cold water tank, taps, ballvalves and vent pipes. In some cases this is impossible to arrange so an air vent may be the best answer. By careful positioning on the

cold water storage tank. The vent tube need only be 15mm bore.

If you want to use this system on pumped central heating circuits you should seek professional advice on the positioning of the tube since it may produce an adverse affect. **Central heating pumps:** If you hear a lot of noise coming from your central heating pump it may be air that is causing it. This only occurs on pumps mounted on horizontal pipe runs.

Turn the system off and look for a bleed screw on the uppermost part of the pump. Undo the screw by about half a turn until water seeps out then tighten it before turning the systems on again.

If the problem is pump vibration try fitting pipe brackets either side of the pump. On variable speed pumps you may stop the noise by turning the speed down. You must turn off the system before doing this.
Pipe sizes: The accurate sizing of pipes is crucial in achieving silent plumbing. Full calculations are complex, especially in heating systems, but a rule of thumb guide—set out below—may help.

1 *Disconnect the elbow joint next to the cold water cistern*

2 *Replace the elbow with a tee joint, one half of the 'T' pointing upwards*

3 *This air chamber is simply made from 28mm tube and a blanking cap*

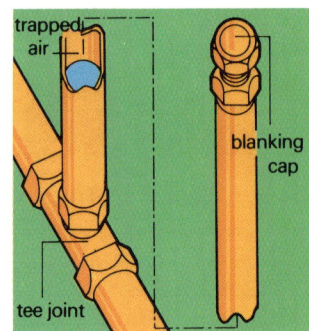

The principle—and the practice—of an air chamber is very simple

By continuing the pipe run above the tank you can produce an air chamber to cushion the shock waves of hammering. Make the piece of pipe as long and large as you can sensibly manage, the only restriction being that it must remain vertical. As an example, 600mm of 22mm copper tube would be the sort of requirement for an air chamber. Cap the top with a blanking-off fitting; tighten it properly and seal it with jointing compound. When you fill the system air will naturally collect and become trapped in the vertical pipe. This air will absorb the shock waves.

After a while the air may become displaced by the pressure of the water. You can recharge the chamber by turning off the main stopcock, draining the pipe through

pipe air will rise towards the valve and be omitted through the top.

If you choose the manual type of air cock operated by a radiator bleed key be sure to run a foot or two of pipe vertically up before fitting the cock.

This will act as an air bottle to collect the air in greater volumes before its release is necessary.

If you would rather fit an automatic version there are several makes on the market. They are designed to go on emitting air indefinitely. Their use is restricted to removing air from pipe runs and they are no substitute for a boiler or cylinder vent.

A third way of removing air from pipe runs is to provide additional vent pipes at high spots. This is done by fitting a tee to the pipe and running a vent tube up to the

Gravity flow and return pipes to and from the boiler should be a minimum of 28mm in diameter. Any less will give rise to a noise resembling a boiling kettle.

Pumped circuits should be a minimum of 22mm in diameter when serving more than three radiators, after which they may be reduced to 15mm on short runs. Bore sizes of less than 15mm must only serve individual radiators.

The cold feed to the cylinder should ideally be 28mm, and definitely not less than 22mm in diameter, especially where the base of the cold water supply is less than one metre above the top of the cylinder.

The hot supply should be 22mm in diameter reducing to 15mm to serve a sink or basin. Indirect cold water supplies must be 22mm.

CENTRAL HEATING REPAIRS

Dealing with cold radiators

If one or more radiators are cold, or significantly cooler than the rest, an airlock caused by air getting into the system or the presence of gas due to a chemical reaction between air, water and metal may be responsible.

Simple airlocks can be removed by bleeding using a special key to open the bleed valve on the top of the radiator.

Before you start, turn the system off and give the air a chance to settle. Hold a container under the valve to catch any drips and then open it very slightly with the key. If there is an airlock, you should hear a hiss of escaping air which will be followed by a dribble of water which gradually becomes a steady flow. When this happens, close the valve immediately.

Occasionally, because of leaks, lack of water in the system, or a design defect in the pipework, the airlock may be more serious. In this case bleeding the radiators could have no effect.

Open the bleed valve on the affected radiator and get one or more assistants to stand by with containers to catch the drips.

Find the pump (normally near the boiler) and locate the flow regulator. Note what setting it's on and then, using a screwdriver, turn it full on and then off in 15 second bursts.

If you still have no success, the only other thing to try is bleeding the pump itself.

You should find the bleed valve on top of the pump casing: open and close it again very briefly using a screwdriver—a hiss of air betrays the fault.

If the radiator valve itself has seized up, drain the system and dismantle the valve. Clean the jumper gently with wire wool and reassemble it; if you can't get a replacement rubber O-ring use PTFE tape instead.

Note that you'll almost certainly get airlocks if you've had to drain and refill the system. But you can minimize them by adopting the correct procedure.

However, if you are constantly plagued by airlocks or you have to bleed a radiator every week, suspect another fault. Check that the expansion tank is at the right level: if dry, the ball valve has probably jammed.

Balancing: Sometimes, what seems like an airlock in a radiator is in fact caused by an incorrectly set lockshield valve. These balance the flow between one radiator and another and are preset during installation. But if the radiator has been removed for any reason, the setting could have been disturbed.

If you suspect that this might be the case, remove the cover and with an adjustable spanner open or close it a few turns to admit hot water to the radiator.

Most central heating systems will look like this. Pipework problems are less common than pump and radiator faults, though old age and hard water take their toll of pipes, radiators, valves and boilers. While the majority of problems are minor ones and easily dealt with, prevention is always better than cure

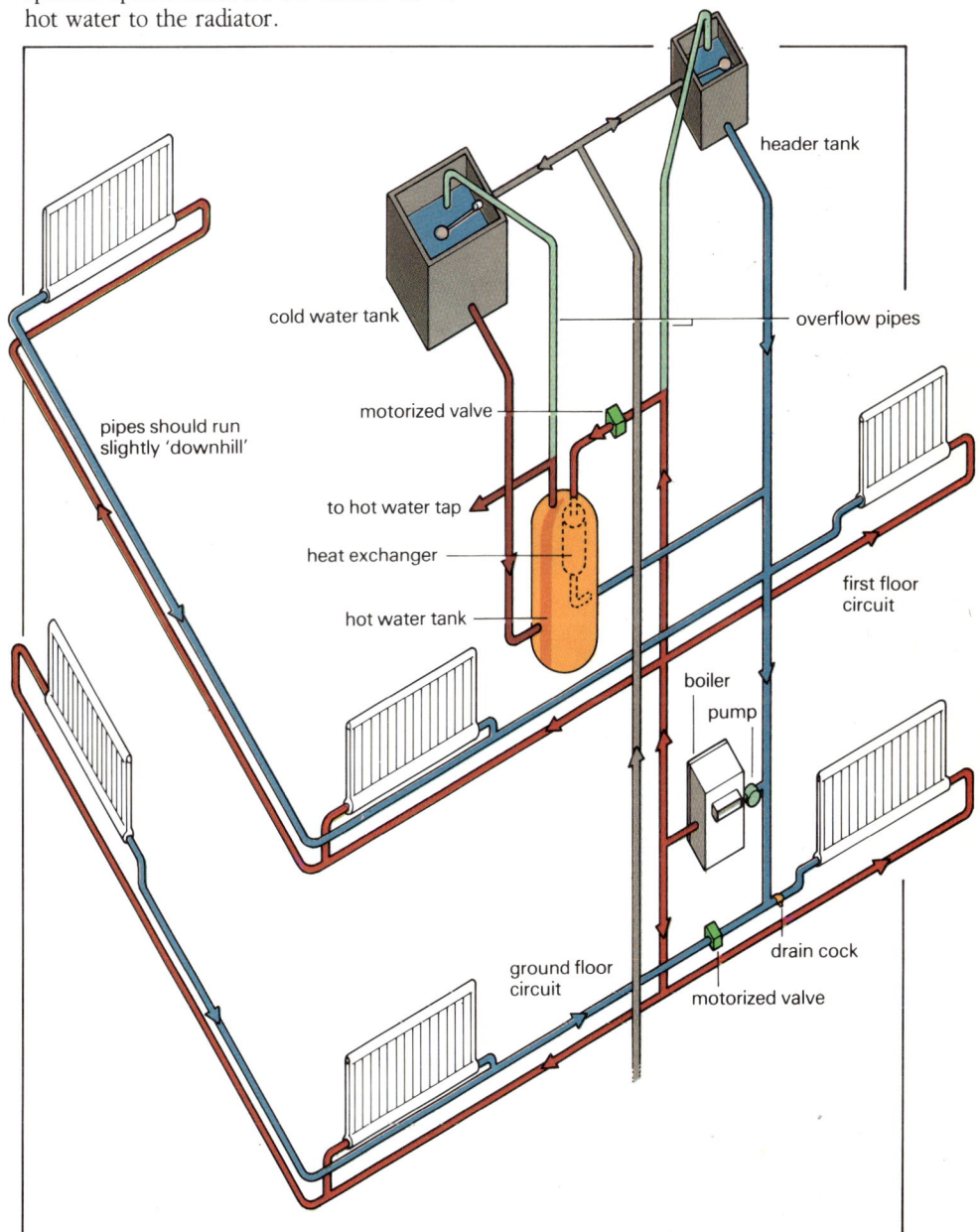

header tank

cold water tank

overflow pipes

motorized valve

pipes should run slightly 'downhill'

to hot water tap

heat exchanger

hot water tank

first floor circuit

boiler

pump

ground floor circuit

drain cock

motorized valve

Faulty thermostats and time-switches

When all your radiators go cold—or get hot when you don't want them to—the first thing to check is your room thermostat.

Make sure it is on the right setting—15°C in a hallway, 21°C in a living room. If you turn the dial back and forth, you should hear a click as the thermostat operates: no click means that it is faulty and should be checked or replaced.

Check, too, that the thermostat isn't giving a faulty reading because of some outside source: a strong draught, a fire nearby or sunlight can all affect it.

If all is well, move to the programmer/timeswitch. Start by checking that it shows the correct time: a momentary power cut could have zeroed it while you were out.

Next make sure that the programme settings are correct—they're easy to get wrong or knock out of adjustment.

If the programmer seems 'dead', turn off the power supply and unscrew it from its backing box. Check that the power 'in' connections are sound and that plug-in connections between the unit and backing box haven't become dirty or loose.

If turning the programmer to 'constant' has no effect on the boiler or pump, check the connections to these two components—again making sure the power is switched 'off'. Loose terminals or frayed flexes should be easy to spot: if necessary, replace the defective wires using heat-proof flex.

If your system includes thermostatic control valves on the hot water primary circuit supplying the cylinder, or motorized valves to distribute boiler-heated water may be able to dismantle it and free the temperature-sensing bellows controlling the flow, but drain the system before doing so.

It may be stating the obvious, but check the boiler thermostat as well. Instructions on the boiler unit will tell you the correct setting, while you can check it's working by turning the dial and listening for a click. Check the sender capillary between the thermostat and heat exchanger—it mustn't be kinked or dislodged. Replace or renew it and test the system.

Boiler and pump faults

A faulty circulation pump may result in cold radiators, over-heating or an excessively noisy system.

You can tell if a pump is running by switching the programmer to 'constant'

1 *Check all wiring and settings, especially in central heating programmers*

2 *The connections in motorized valves can vibrate and so loosen*

3 *Don't forget to check thermostatic radiator valves too*

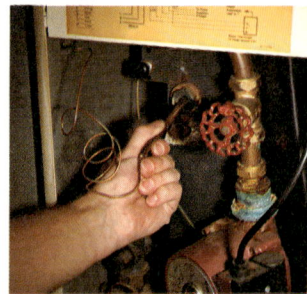

4 *Check the sender capillary is located correctly and not kinked*

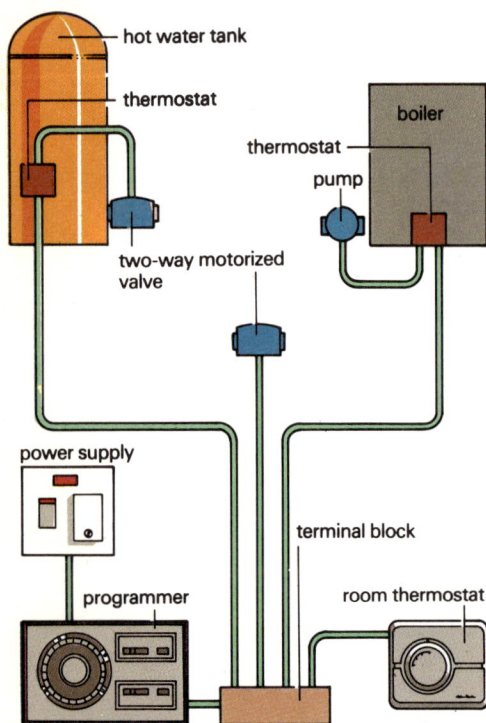

between the hot water circuit and the radiator circuit, you may experience trouble due to jamming or failure of the motors.

The most common symptom, assuming the programmer is working, is that either the hot water or the heating valves don't work when they should.

You'll find such valves near the boiler of the cylinder. With the system switched to 'constant', feel by hand that hot water is being distributed through them.

On a motorized valve, check that the electrical connections to the motor haven't become loose.

On some systems, thermostatic valves are fitted to individual radiators to give localized heating control. These are mechanical in operation, and sometimes they may jam.

Depending on the make of valve, you

Electrical connections and cable runs are basically simple; keep wiring diagrams and maker's instructions handy to deal with problems quickly

and holding a screwdriver to it, like a stethoscope: if you hear a noise, suspect some other fault—like an airlock; if there is nothing, either electricity isn't reaching the pump, the motor's burnt out, or the pump rotor is jammed.

Assuming that you've checked the power supply and programmer, the fault lies with the pump.

On some pumps, such as the common *Commodore*, it's possible to check the rotor without removing the unit, providing stop valves are fitted either side.

Shut off the power supply and close the stop valves on either side of the pump by turning them clockwise. Loosen the screws securing the casing and remove.

The rotor underneath should spin freely: if it doesn't, you may be able to free it by *gentle* levering with your screwdriver.

In all other cases the pump must be removed and taken to a heating suppliers to be checked.

If there are no stop valves, the system must be drained completely before removal.

1 *Listen for the sound of the pump motor turning by touching the blade of a screwdriver to the body*

2 *Use the valves on either side to isolate the pump from the water supply before removal*

3 *Check the connections inside the pump itself. Some of them may well have shaken loose*

4 *When you re-fit the pump, replace the O-ring or hemp seals on the coupling joints*

The pump will probably be connected to the pipework by threaded screw couplings: loosen these with a pair of plumber's wrenches or adjustable spanners and it should come straight out.

Before removing the pump completely, you must sever the electrical connection to the programmer. On some pumps you can get to the terminals easily having first removed a small cover on the motor casing.

With other types the motor is sealed, so disconnection must take place at the programmer end.

Fitting a new pump is a straightforward reversal of the removal procedure. But be sure to seal any screw coupling joints with hemp and jointing paste before you finally tighten them up.

Certain types of connector (especially those which incorporate a shut-off valve) may not require hemp; instead, replace the rubber O-ring seals.

★ WATCH POINT ★

Before you disconnect the wires make a note of which goes where and label them for reference.

Corrosion

Modern central heating systems in which the radiator and hot water cylinder primary circuit water is constantly recirculated should not suffer from scaling.

In areas with exceptionally hard water it is possible that the secondary hot water circuit could be affected.

Some corrosion in a system is perfectly normal—it's what turns the water black. Most systems have a proprietary corrosion inhibitor added to the primary circuit water via the expansion tank: replace it when you

---

Faultfinder chart

If your central heating goes wrong, consult this chart first—it should help you get straight to the root of the trouble

One radiator cold:
Check valves
Bleed radiator
Check for air locks
(see Dealing with Cold Radiators).
Check radiator thermostat
(see Faulty Thermostats and Time-switches).

Overheating:
Check boiler thermostat
(see Faulty Thermostats and Time-switches).
Check pump (see Boiler and Pump Faults).
Check water supply (see Dealing with Cold Radiators).

All radiators cold:
Check room thermostat and time-switch/programmer
(see Faulty Thermostats and Time-switches).
Check boiler (see Boiler and Pump Faults).
Check pump operation (see Boiler and Pump Faults).
Check water supply (see Dealing with Cold Radiators).

Excessive noise:
Check pipe/radiator brackets and notching in joists/floor.
Check pump (see Boiler and Pump Faults).
Check for scale and corrosion (see Corrosion).

Fumes in house:
Check boiler (see Boiler and Pump Faults) and ensure flue is not blocked.

---

drain the central heating system.

Excessive corrosion will only occur if large amounts of air are present in the system or the primary circuit water is too

cool. So deal with leaks and airlocks immediately, and set the boiler thermostat above 60°C. This should help to eliminate corrosion in a relatively short time.

MOVING A RADIATOR

There are many good reasons for repositioning a radiator. It may be, for example, that you intend to improve your home by knocking two rooms into one or by replacing a window with sliding patio doors. If this is the case, you may not have any option but to re-organize your central heating layout.

In many houses, radiators were originally positioned to suit the convenience of the installer rather than to suit the occupier. This can mean that a radiator lies along a wall exactly where you want to put a sofa. Or you may just want to exchange an old radiator for a more streamlined or efficient model. Whatever your particular reasons are for moving a radiator, plan your approach to the job with care so that you can avoid potentially messy disasters. Make sure you have plenty of dust sheets to hand.

Planning considerations

The two most important points to consider are the type of system you have and where you intend to put the radiator in relation to the existing layout.

The system: The pipework which connects the radiators to the boiler will be **smallbore** (15mm or 22mm in diameter) or **microbore** (6mm, 8mm or 10mm).

Smallbore pipework will follow one of two different types of layout. In a one-pipe layout, a single 22mm pipe runs around the house to and from the boiler and each radiator is joined to it via two connections. However, the majority of houses have a two-pipe layout where there are separate pipes for flow and return with the radiators connected to bridge across them (see diagram on this page). In a two-pipe layout, more than one radiator can be connected to the bridging branch.

With a two-pipe smallbore system, the size of pipe decreases as it gets further away from the boiler—it starts off with 22mm and reduces to 15mm, whereas in a microbore set up, the pipes connect the individual radiator to a centrally placed manifold.

In order to identify exactly which system you have, you will inevitably have to lift a few floorboards so that you can trace the flow and return pipes back to their source.

Choosing the new position: Unless you have double glazing—or plan to install it—it's best to put the radiator directly underneath a window. This is so that the heat from the radiator can counteract the cool downdraught from the window.

If you have double glazing, put the radiator where it is most convenient or where you will have to use the least amount of extra piping.

With smallbore systems, you may be able to use the existing branches if the new site is near to the original one—it's simply a matter of extending the pipes with the aid of elbows or straight connectors.

If the new site is some way from the old one, you'll have to break into the main flow and return pipes (two-pipe layout) by installing tee fittings and connecting new branch pipes. The old pipes can be blanked off either at the old radiator position or where you tee off (you can re-use the old pipes if they are in good condition).

Once you have decided where to put the radiator and where to break into the circuit, work out what you need to buy in terms of tees, elbows and new piping. To help in your calculations, it's a good idea to make a scale drawing of your room. If you are connecting metric sized pipe to imperial, 15mm compression fittings will fit both pipe types but 22mm won't, so in this case you'll have to use a special capillary adaptor to make the join.

With a microbore system, you have the additional choice of re-routing the pipes

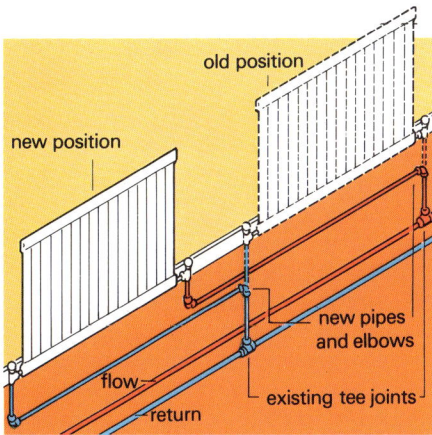

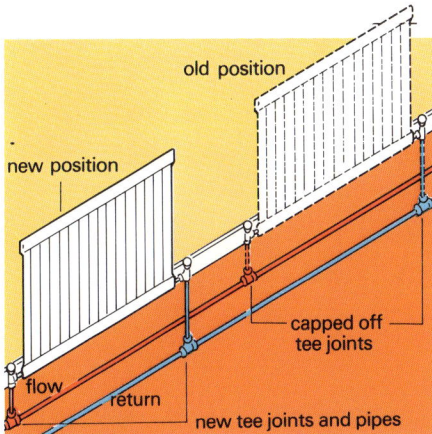

With smallbore systems, use whichever of the methods shown above is most convenient to run the pipework to the resited radiator

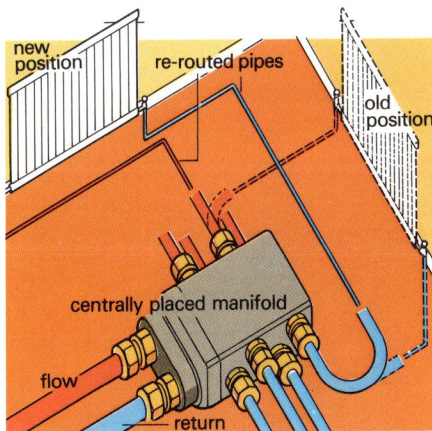

With microbore systems, each radiator is connected to a centrally placed manifold so run your new pipework back to that

from the manifold. This may be easier than trying to connect to the existing pipes. Microbore is notoriously thin and fragile and unless you are used to handling it, you will be better off buying a new coil of pipe which will be much easier to work with than the old piping.

If your microbore supply pipes are to be clipped to the wall or skirting (as opposed to running underneath the floor), a good way of protecting them is to encase them in plastic conduit. Cable conduit is adequate but check that it can take the diameter of the pipe.

Tools and equipment: For removing and fixing the radiator, you will need little more than a screwdriver, an adjustable spanner, some jointing compound and some PTFE tape. However you will also want tools for cutting and jointing pipe, lifting floorboards and drilling holes through joists. A garden hose will prove invaluable when it comes to draining down the system you can carry waste water from the house without spillage.

Removing the radiator

Before you start disconnecting the radiator, turn off the boiler and let the water cool down. Then drain the system.

To drain the system, turn off the valve in the supply pipe to the feed-and-expansion cistern in the loft—this is smaller than the main cold water tank. If there isn't a valve, tie up the ballcock.

Next, find the draincock which will be located at the lowest point in the system, probably next to the boiler. Push one end of a garden hose over the nozzle on the draincock and lead the free end to a gully or drain outside the house. The level of the drain must be below that of the draincock.

Turn off all the radiators before opening

the draincock and letting the water flow out.

When the pipes are empty, make sure that the handwheel valve on the radiator you want to move is tightly shut and close down the lockshield valve with an adjustable spanner. As you twist the lockshield nut, count the number of turns necessary to close the valve as you'll need to open it by this number of turns later.

The next step is to undo the two compression nuts which connect the supply pipes to the valves. Residue water will probably come out of these as you loosen them, so pull any carpet away from under-

2 *Undo the compression nuts which secure the supply pipes to the handwheel and lockshield valves. Be prepared for leaks*

1 *To drain the system, find the draincock, push a garden hose onto its outlet and then open it with an adjustable spanner*

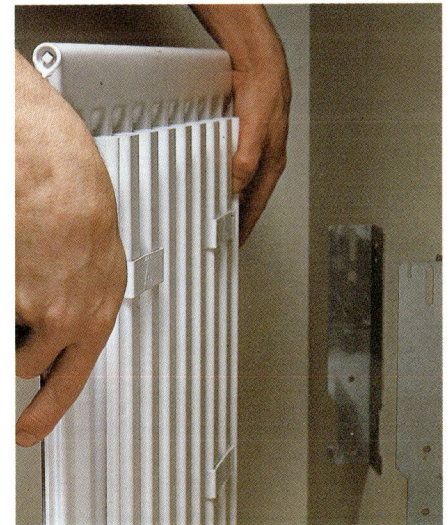

3 *Free the radiator from the pipes and lift it clear of its mounting brackets. Remove the brackets and make good their fixing holes*

4 *If you're going to re-use the radiator, take the opportunity to flush it through with a strong burst of water from your hose*

5 *Cut back the supply pipes to the radiator in its old position so that you're ready to run the new piping in later*

selves, play a gentle flame onto the radiator side of the joint for a few seconds.

Hanging the radiator

Hang the radiator in its new position before you lay down the connecting pipework— that way you will know exactly where to lead the pipes.

A new panel radiator has four screwed tappings—one at each corner. Only three of these are used—the remaining one is blanked off. Normally, the flow and return pipes are connected to the bottom two tappings and an air bleed valve screwed to one of the top tappings.

6 *If you're upgrading as well as resiting a radiator, measure up for the fixing brackets and transfer your measurements to the wall*

8 *Fit the brackets. Now is also the time to fit heat reflecting foil especially if you're installing the radiator on an outside wall*

You need to fit the brackets to the wall so that the bottom of the radiator is just above the skirting board. If you are siting the radiator underneath a window, you also need to leave a 50mm clearance below the sill. To site the radiator properly, lay it on the floor and clip on the brackets. Measure and make a note of the exact distances of the brackets from the bottom and sides of the radiator. Transfer these measurements to the wall and mark where to drill fixing holes.

Before you screw the brackets to the wall, check beneath the floorboards for joists. If necessary, shift the radiator to one side so that the joists won't obstruct the connecting pipes. A second consideration—if you are fixing to a timber framed wall—is the position of the studs. If you can't screw

7 *With the radiator fitted to the wall and the valves fitted to the radiator, mark on the floor the entry points for the pipework*

9 *If you've invested in a new radiator, it's worth fitting a thermostatic valve as it will give you greater temperature control*

neath the radiator and put a bowl below the nut. Free the radiator from the pipes and lift it clear of the brackets.

If you are fitting a new radiator and want to salvage the old lockshield and wheel-valves, now is the time to remove them. However, some old radiator valves are not compatible with their new counterparts, so check before you start setting to with a spanner. If the nuts prove very obstinate—which is likely to be the case if the radiator has not been handled for a long time—try heating the joint with a blow-torch. So as not to damage the valves them-

directly into wood, mount a horizontal batten between studs and secure the brackets firmly to this. Large radiators may require three brackets.

Once you are satisfied, drill holes in the wall, insert appropriate wallplugs and screw the brackets home using a quantity of 50mm No. 10 screws.

Hang the radiator on the brackets and, if

> ### ★ WATCH POINT ★
>
> If they are in good condition, there's no reason why you can't re-use all the original fittings, but this is a good opportunity to fit a thermostatic wheel valve.

you have a wooden floor, use an offcut of pipe as a guide to mark holes directly underneath the two valves. Drill clearance holes through the floorboards for the pipes.

Running the pipes

The amount of work you will have to do will depend on how far you are moving the radiator. It's a good idea to lay all the pipes and fittings out in a dry run—from the connection points to the radiators—before you start connecting up. Remember to run pipes parallel with, or at right angles to, the joists.

Cut the pipes to length and when you are happy with the layout, start joining them together—see Handling copper pipe. Leave the final connection to the radiator until last—the flow pipe should lead to the hand-wheel valve.

On a solid floor, chase the pipes into the wall above the skirting. Running pipes underneath a suspended wooden floor is not so easy and you will have to lift floorboards to gain access. Downstairs, all the pipes can be run below the level of the joists; upstairs you'll have to cut some slots in the joists to take pipes running at right angles to them.

Microbore pipe can be pushed through 10mm holes drilled through joists in much the same way as electric cable but, with smallbore, you'll have to cut a slot to take the pipe. To protect pipes that run through joists from floorboard nails, either screw metal plates along the top of the joists or fill each slot with a vee-section cut-out.

Secure the pipes with plastic clips screwed to the sides of the joists (or to the wall if there's a solid floor) at 1m intervals.

Make the connections to the radiator by leading the pipes through the holes in the floorboards and up to the valves. Be sure to apply jointing compound to the end of the

pipe before tightening up the compression nuts. Don't be tempted to over-tighten the nuts as this will crush the pipes and weaken the joint, and may cause leaks. If joints don't work it is best to separate and re-make them.

Re-filling the system: Close down the draincock and double check that you have capped off any severed pipes. Open up the wheel valve on the radiator and twist open the lockshield valve the same number of turns that were needed to close it down when you started the job.

Open up the supply valve to the feed-and-expansion cistern (or untie the ballcock). As the water starts to flow through the system, go round the house bleeding all the radiators to get rid of any air. If you turned off all

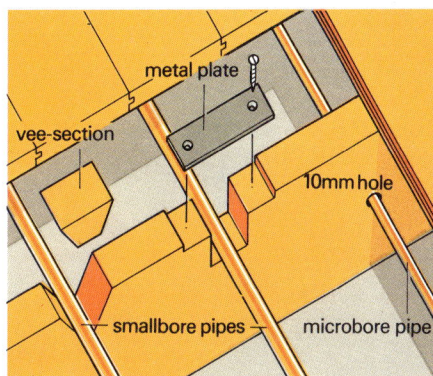

Always protect pipe when running it at right angles to joists. Microbore can be run through joists, smallbore must be notched into them

the other radiators when you drained the system, they should still be full of water but it's best to make a thorough check and to bleed them as well.

Bleeding is usually done by turning the bleed screw with a special key and letting the air out—as soon as water spurts out of the hole, you know that the radiator is full. Some modern radiators have specially designed air vents which are opened and closed by hand—thus making the use of a key obsolete.

As the system is filling, check carefully for leaks and tighten any suspect connections. It helps to have someone in the loft who can turn off the water quickly if you discover a major leak.

When the radiator and system are full, switch on the boiler and nail all the floorboards back into place.

Should the re-filled system start vibrating (known as 'knocking') when you turn on the heating, don't despair as the chances are that an air lock will be the cause. The

simple cure for knocking is to bleed all the radiators once again; if necessary, bleed the valve on the electric pump too.

Handling copper pipe

It's easiest to cut copper pipe with a special pipe-cutting tool. This has a hardened steel wheel which scores a line around the pipe as you rotate the tool. The score gradually gets deeper and eventually slices through the copper. Use the pointed bit on the end of the tool to remove any burr you've made.

When cutting into a length of existing pipe to make a tee-joint, remember that the section you cut out is less than the overall length of the tee—the shoulders on com-

10 *When you know where you want to run the new pipework, screw pipe clips to the joists at 1m intervals*

11 *Run new piping to the point where you cut back the old, adjust its length and then make the connections*

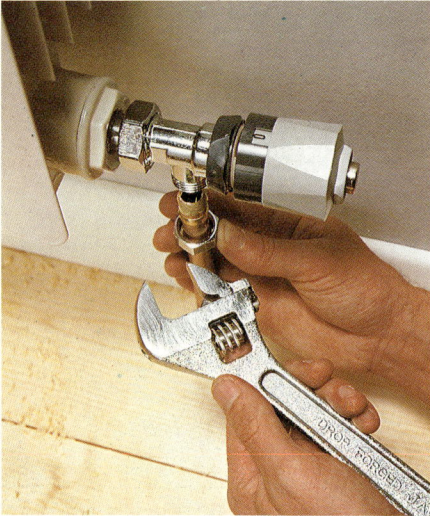

12 *Bring your pipework up through the floor and make the connections to the radiator valves—be sure to use jointing compound*

13 *Tighten the lockshield valve. With older radiators you'll need an adjustable spanner, but newer types may need a screwdriver*

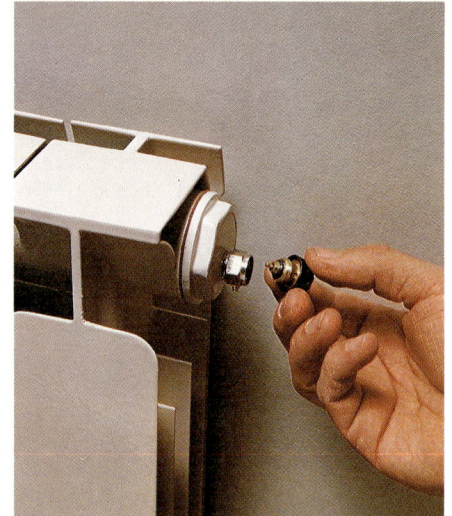

14 *The final stage of the job is to bleed the radiators. When water spurts from the bleed valve, turn the key or replace the nipple*

pression tee fittings show you where the pipes should end. You will have to flex the supply pipes apart so you can slip the fitting into place.

When it comes to joining pipes together or capping off severed ends, compression fittings are the easiest type to use—especially in confined spaces.

To make a compression joint, first slip the capnut and olive over the end of the pipe and then wipe on some jointing compound. Push the pipe into the fitting until it butts up against the end of the recess. Twist on the capnut until it is hand tight and then give it a final half turn with a spanner—it's important not to overtighten the nut.

Scrape away any burr on the inside of the pipe using the point on the pipe-cutter

A pipe cutting tool is well worth hiring if you've got a lot of joints to make

Compression fittings are the easiest to use. Smear plumber's putty on the pipe end first

HOW TO CLEAR BLOCKED DRAINS

How drainage systems work

Before you attempt to unblock main drains, make absolutely certain what system you have and get to know how it works.

Two pipe systems are still very common on houses built before the Second World War. There are two separate waste stacks running down the outside of the house—one for waste water and one for soil (waste from the WC). The waste pipes from your plumbing fittings run into the waste stack either directly or via a hopper head

Below: typical single stack system with an extra soil pipe (plus rodding eye) joining the main flow at a manhole

Right: older two pipe system with hopper on the waste stack. The combined flow is joined by a gully at the interceptor

(now obsolete, but still very common). Pipes from ground floor fittings often connect to the stack underground. But if they are far away from the stack they run instead to a separate gully—a kind of underground U trap. This joins the underground pipe from the waste stack at an inspection chamber, covered by a manhole.

Soil from the WC always runs to the soil stack direct. The underground pipe from the stack joins the waste water pipe at the inspection chamber.

Rainwater may be collected at a gully to join the waste water system. It may run from the gully to the inspection chamber via a separate pipe. It may be dispatched to a separate gravel-filled pit or soakaway. Or, in areas where water is in short supply, it may run to a separate stormwater drain.

From the inspection chamber, the combined waste and soil water flows toward the main drain, normally in the road. Before it gets there it may well pass through another

chamber—the interceptor—containing a large U trap.

Interceptors were once used to cut houses off from the main drains; this is no longer done, so they are no longer fitted. But you may find that you share an interceptor with one or more neighbouring properties Interceptors are easily distinguished from ordinary inspection chambers by an air inlet terminal nearby.

The single stack system is the one now in common use. As its name implies, waste water and soil pipes all connect to the same stack. Until recently the stack had to be inside the house, but the rules have since been relaxed to allow outside stacks. Ground floor appliances too far away from the stack to connect to it have their own sub stack or run to a closed (back inlet) gully. All underground pipes run in a straight line to meet at an inspection chamber.

Variations: There are as many of these in drainage as there are in plumbing. Houses

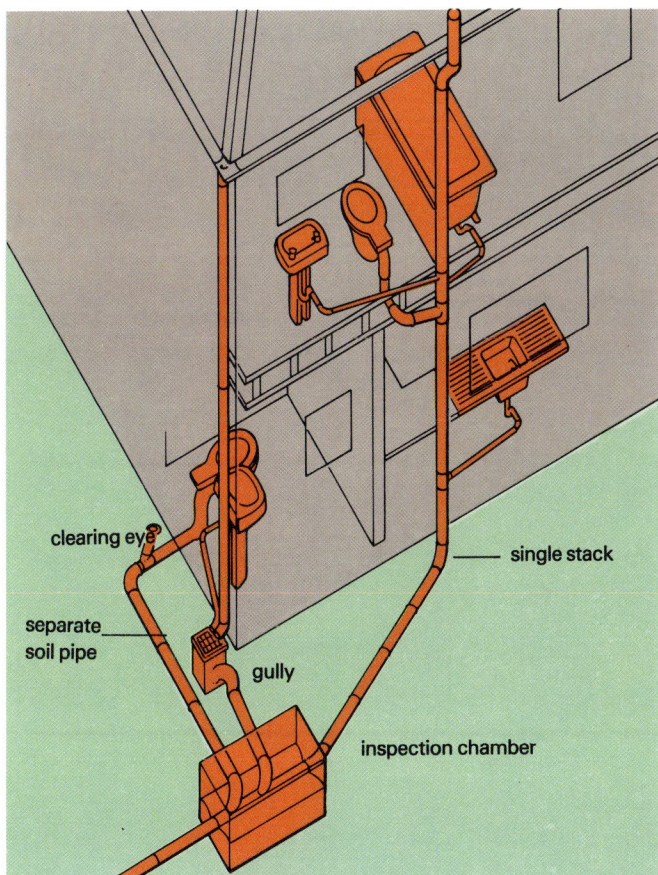

clearing eye

single stack

separate soil pipe

gully

inspection chamber

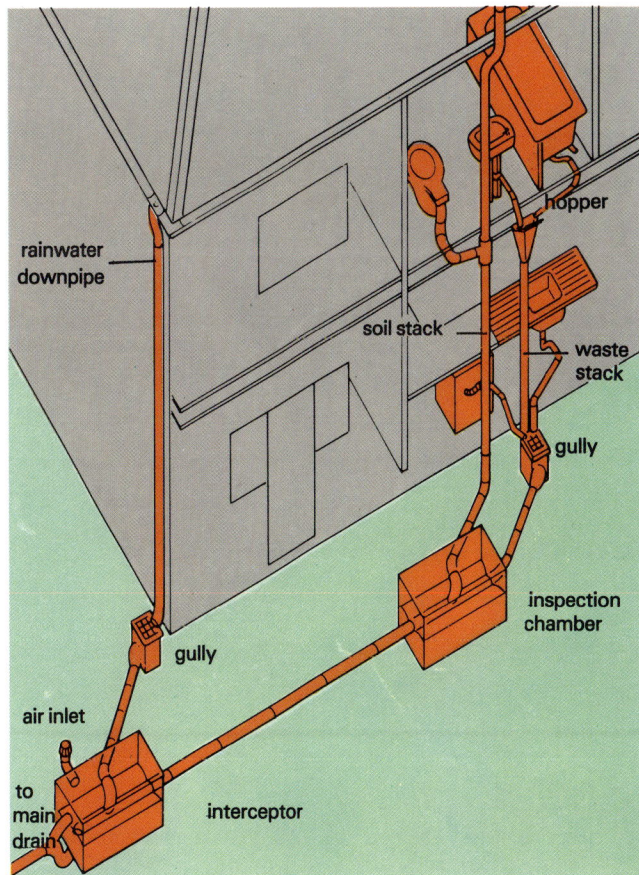

rainwater downpipe

hopper

soil stack

waste stack

gully

inspection chamber

gully

air inlet

to main drain

interceptor

with the two pipe system which have been modernized may also have an internal single stack or sub stack.

Some larger houses have the **one pipe system**, in which a single stack runs on the outside of the building.

Some early single stack systems have additional old-style gullies.

And you may find you have more than one inspection chamber: they must be installed wherever pipes join and where the gradient or direction of the drain changes.

The only way to be really sure how your drains are laid out is to piece them together on a sketch plan, using the information above as a guide.

Locating blockages

Your first check is always to see if the trap on the fitting itself is blocked. If you are sure the blockage is in the drains, adopt the following procedure.

Open the manhole nearest the house. If this is empty the fault is in the fitting waste pipe, the stack or the gully trap (in the case of a backed-up gully).

If only one fitting on a stack gives trouble, the fault is in its own waste pipe. If several

If the handles are intact, use timber and rope to lift cover

do, the blockage is in the stack—probably low-down, towards the chamber.

If the manhole is full of effluent, don't immediately assume that this is where the blockage is. Try if possible to lift all other manholes between there and the main drain in the road—particularly if the drain includes an interceptor, which give more trouble than ordinary inspection chambers. It may be that the last one in the line has blocked, causing effluent to back up as far as the first.

Lifting a manhole cover

You have to lift the manhole cover to check the drains, so this is always the first job. It's not easy: cast iron manhole covers are heavy, so get some help.

Frequently the cover is rusted in its frame. Scrape around the join with a screwdriver then tap the cover gently with some wood—the vibration should be enough to release it from its seating.

Special keys are available for lifting covers but if you don't have one a strong hook or a piece of steel bent in a vice will do the job just as well.

Some covers have handles consisting of small bars across indents. In this case loop several turns of string or wire through the bars and around stout pieces of timber. With someone on either side of the cover and using the timber as handles, lift the

cover and swing it free. If necessary, as a last resort, use a garden spade to lever up the cover over a fulcrum made from wood or bricks. The easiest way is to lift one end first and support it across the opening on a broom handle or similar. You can then lift the other end onto another support and use them both as rollers to push it clear of the opening.

Some covers are secured by screw bolts, in which case soak them in penetrating oil before attempting to undo them. Remove the bolts with a spanner or wrench.

On all manhole covers a little grease smeared around the frame before replacement will stop future rusting.

Blocked inspection chambers

The only really effective tools for this job are a set of rods. You could try plunging with a mop or poking with a bamboo stick, but because the outlets and inlets will be hidden by effluent, you really need something more flexible.

Drain rods can, of course, be hired, but this is inconvenient. If your drains block regularly or you are worried that they could, it really is worth buying a set of 'unbreakable' polypropylene rods and attachments—they are now very cheap. Among the attachments you can get are a plunger (for gullies), a worm screw (for retrieving debris), a scraper (this unfolds when you withdraw the rods), a flexible wire leader (for acute bends) and a small wheel (this screws onto the front rod to stop it catching on joints).

Screw two rods together and lower them into the chamber. Keeping the diagram in this panel in your mind's eye, feel for the inlet or outlet where you think the blockage is, and try to slide the rods into it.

When you feel them go in, screw on

1 *Lay out your rod attachments before assembling the first rods*

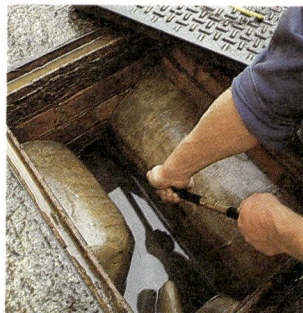

2 *Try the plunger head first, working it vigorously*

3 *After the blockage has cleared hose the chamber thoroughly*

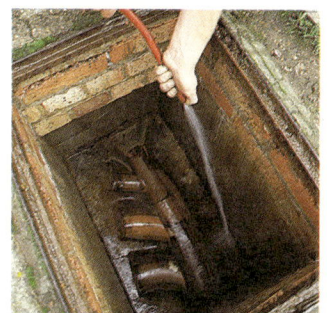

4 *Chambers vary in size, but all demand the same procedure*

another rod and keep pushing. Continue in this way until you feel an obstruction.

Now turn the rods clockwise and at the same time keep manoeuvring them back and forth until you feel the blockage 'give'. You may have to do this for several minutes—and quite vigorously—before it has any effect.

After you have cleared the blockage, always flush the chamber thoroughly with water to remove the last traces of debris.

Interceptor problems

Interceptor blockages are nearly always in the trap. Above this is another pipe—the clearing eye—which should be closed by a clay stopper on a chain.

When confronted with an interceptor full of effluent, the first thing to try is removing the stopper. You won't be able to see

stopper chain breaks and the stopper falls into the trap where it acts like a butterfly valve and causes periodic blockages. If you think this has happened, call in professional help. If the stopper is simply missing from your interceptor, you can buy a new rubber one from a plumber's merchant.

Clearing gullies

Gullies, particularly the old open sort, are very common sources of blockages. If the blockage is in the trap, you can usually clear it from the gully itself. But if all else fails, rod in a plunging motion up the inlet in the inspection chamber that leads to the gully.

Small gullies can be baled out by hand. Lift the grid, then, wearing rubber gloves, scoop out the debris into a bucket. Check that the outlet is clear with your finger: if it isn't try poking some coathanger wire up it.

Gullies receiving sink waste get clogged with grease. This can usually be shifted with boiling water or scraped out by hand.

If this doesn't work try pouring down a solution of caustic soda. But take care and follow the instructions to the letter—caustic soda is dangerous.

Deep gullies and blockages on the other

gully itself with polyethylene bags rammed tightly in place—the extra pressure this creates will help the water do its work.

Blockages in stacks

These are rare in single stack systems and simple two pipe arrangements, but common in older, complicated drainage layouts. Fortunately there are often clearing eyes at junctions and direction changes.

The procedure is to soak eye covers (or their bolts) in penetrating oil before attempting to remove them.

You can then either insert a rod with a head to match the stack diameter, or else feed in a screw clearing wire. Use the same procedure as for an inspection chamber.

Hopper stacks can be cleared from above with a worm screw rod.

Tree roots

These can work their way through a joint in a drain pipe, in which case no amount of plunging will shift them. You may be able to clear them partially with a powered screw augur similar to the type used by pro-

Some gully pipes have rodding eyes at changes of direction. Unscrew the cover for access

Some waste pipe clearing eyes are unscrewed with a wrench. Soak in penetrating oil if stiff

Feed in a screw clearing wire to unblock the waste pipe. Use the same procedure as for an inspection chamber

Hopper stacks which are blocked up can usually be cleared by hand or with a worm screw rod

it—you must feel for the handle with a piece of wood and try hooking it out of its socket.

If you succeed and the blockage is in the trap, the chamber will clear and you can set about clearing the trap itself. Plunge it vigorously with your rods—preferably using a plunging head—until the debris shifts.

If removing the stopper has no effect, the blockage is further down the drain; in this case rod through the clearing eye.

An all too common problem is that the

side of the trap can be shifted by plunging—you can get an attachment like a sink plunger that screws onto the end of an ordinary drain rod. A household mop makes an effective standby, as do old rags wrapped in a plastic bag and tied to a broom handle, but only if there is enough bulk to fill the gully neck completely.

You could try threading a garden hose around the bed in the gully trap to flush the blockage with water. In this case stop up the

fessional drain clearers—they are available at some hire stores. But a more satisfactory solution is to excavate the drain.

Find out where this is by making a note of how many rods it takes to reach the obstruction. Then lay the same rods out at ground level and see where they end.

Dig down to the pipe, hack away and pull out the offending roots, then make good the damage to the joint with a strong mix of one part cement and two of sharp sand.

HOT WATER PROBLEMS

Inspecting your system

Hot water cylinders or tanks are invariably fed from a cold water storage cistern in the loft. In flats, some hot water cylinders are fed from an all-in-one cold water cistern that sits directly on top of the hot water cylinder.

The water can be heated in one of three ways: by a direct system; by an indirect system or by an immersion heater. Before replacing your tank, you must know which system is in use in your house, so go into the loft and make an inspection.

In a **direct** system, the domestic hot water is heated by a boiler, back-boiler or gas circulator. Water flows from the cold water cistern in the loft to the base of the hot water cylinder. From there it passes to the boiler, is heated and rises back to the top of the cylinder. Heavier, colder water from the base then flows from the hot water tank to the boiler and the process repeats itself in a continuous loop.

In an **indirect** system, the water in the hot water cylinder is not heated directly by the boiler: there are two separate circuits. The boiler is connected to the 'primary' circuit; the water in this circuit passes through a coil of pipe inside the hot water cylinder (known as the 'calorifier') after being heated by the boiler. This coil passes heat to the domestic hot water which forms the 'secondary' circuit: the two do not mix.

An **immersion heater** can be added to these systems or fitted on its own to a hot water cylinder. It is simply an electric heater, thermostatically controlled, which is immersed in the water inside the cylinder to heat it.

Replacing an immersion heater

The majority of hot water cylinders have a screwed hole, known as a *boss*, to take an immersion heater. There are two main types of heater: **vertical**, which fit into the top of the cylinder and **horizontal**, which fit into the side.

Single and dual element vertical heaters are available. The dual element type has elements of unequal length. This allows you to select, by means of a changeover switch, whether to heat the water in the whole tank or just that at the top (for the occasions

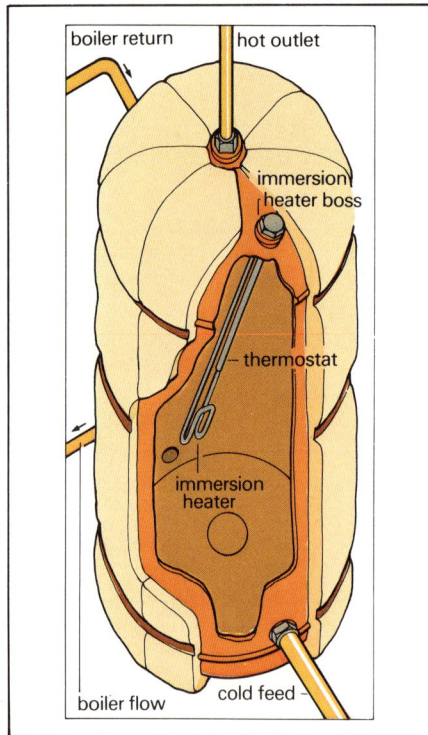

A direct hot water cylinder with an immersion heater

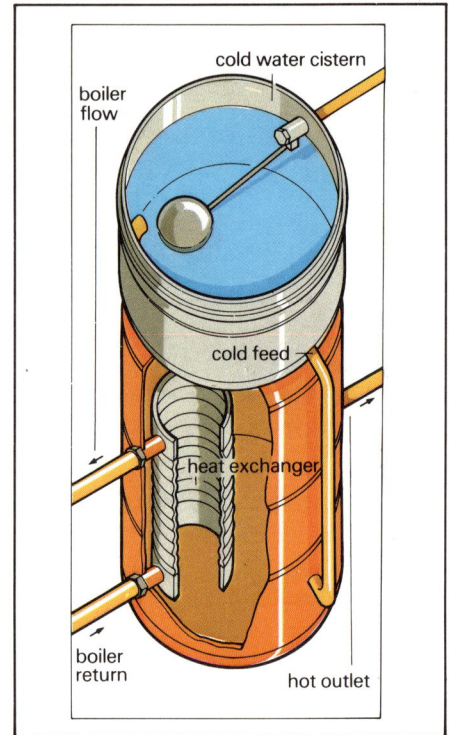

An indirect cylinder with an integral cold water tank

when you only need a small amount of hot water for washing or bathing).

Dual element horizontal heaters do not exist. Instead, you fit two heaters, one to the top and one to the bottom of the tank. If one only is to be fitted, position it at the bottom.

Immersion heaters are generally rated at 3kW and wired on their own circuit, which is protected by a 15 amp or 20 amp fuse. All are fitted with a thermostat which turns the electricity supply to the heater on or off depending on the temperature of the water. Thermostats are often supplied, separately, so make sure you've got one and that it is the right length to fit inside your own particular immersion heater.

Apart from the appropriate immersion heater (and thermostat, if necessary), you'll need some plumber's putty or PTFE tape for sealing the threads of the heater, an immersion heater spanner (these can be hired, but are fairly cheap to buy), a small or adjustable spanner for unscrewing the drain cock and electrical tools (screwdrivers, wire strippers, pliers) for connecting the flex to the heater.

The first steps are to turn off the electrical supply to the heater (switch off the electricity *and* remove the fuse if it's on its

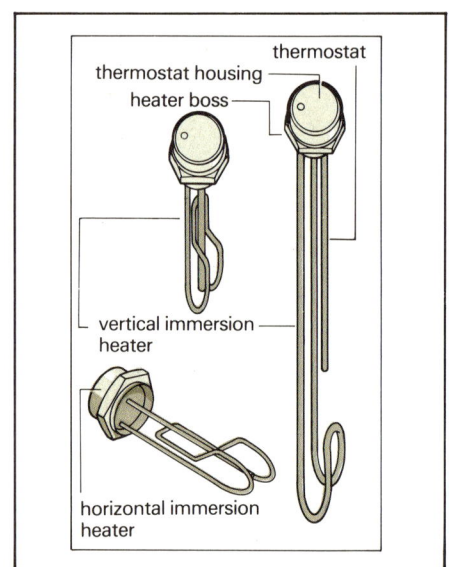

Vertical or horizontal heaters can be fitted to the cylinder

own circuit) and turn off any additional heating supplying the hot water cylinder, such as a boiler or gas circulator.

The gate valve on the supply from the cold water cistern should be closed next—this is likely to be near the hot water cylinder rather than near the other gate valves in the loft.

If there isn't a gate valve, you'll have to drain the cold water cistern after the water has been turned off or the ballvalve tied up to a piece of wood across the cylinder. Fit a gate valve before refilling the system.

Once the gate valve has been closed (or the cold water cylinder drained), connect a piece of hose to the drain plug (positioned next to the cylinder or by the boiler if you have a direct hot water cylinder) with the other end run to an external gully, and open the plug gently with a spanner—don't use pliers which will chew up the plug.

You don't need to drain all the water out of the cylinder—only enough so that water won't come out when you remove the immersion heater. The actual amount will depend on whether your immersion heater is fitted at the top or bottom of the cylinder. It will not be necessary to open the hot taps to drain the system, but it might help to get rid of water in the vent pipe.

The existing heater can now be disconnected from its electrical flex and unscrewed carefully from its boss with an immersion heater spanner.

Once the heater is screwed firmly in place, it can be wired up using three-core heat-resisting flex. The thermostat needs to be positioned carefully inside the heater (after removing the cover) and set to the right temperature before the cover is replaced.

Before refilling the cylinder, all drain taps must be closed and the upstairs hot taps opened to avoid any air locks.

Open the gate valve on the cold supply and watch the immersion heater boss as the cylinder is filling: if there's a leak, tighten the heater until it stops or, if this doesn't work, turn off the valve, drain the cylinder and refit the heater.

If there are no leaks, keep filling until water is running happily out of the hot water taps and then turn these off.

If all is well, the immersion heat can now be turned on. The neon light on the fused flex outlet will tell you that the electricity is connected to the heater—within a short time you should be able to feel the cylinder warming up. Relight the boiler.

Wiring an immersion heater

An immersion heater should have its own circuit from the consumer unit wired in 2.5mm² two-core and earth cable and protected by a 15A or 20A fuse. The cable is run to a double-pole switch from where a length of two-core and earth heat-resistant 13A or 15A (1.25mm² or 1.5mm²) flex connects it to the heater.

A convenient type of double-pole switch would be a switched fused connection unit with a neon indicator, though there are special 20A double-pole switches (with a neon indicator) marked 'water heater'.

If you have a dual element or two separate elements you will need a special changeover switch (often marked 'bath/sink') and two lengths of flex will need to be run to the heater. Inside this type of switch, you will need an extra piece of insulated single conductor cable (take a bit from spare 2.5mm² two-core and earth cable) to connect the two parts of the switch together.

You may want to have a time-switch connected to the immersion heater so that it comes on before you get up in the morning and perhaps before you come home from work in the evening. There are special time-switches for this, wired in as shown below. Don't use a plug-in timer to control an immersion heater—it simply is not robust enough to handle the job.

Fitting the new heater

Before fitting the new immersion heater, smear plumber's putty on the threads (or wrap PTFE tape around them) so that the joint won't leak afterwards.

When wiring the double-pole switch, the cable from the consumer unit is connected to the terminals marked 'SUPPLY' or 'MAINS' and the flex to the heater to the terminals marked 'LOAD'. At the heater end, there are terminals helpfully marked L and N for the brown (live) and blue (neutral) wires and E for the earth wire.

Renewing a hot water cylinder

Houses built before about 1939 may still have an old-fashioned rectangular galvanized steel hot water tank, together with its associated lead plumbing. This type of tank has a circular inspection hand-hole on the front and is not fitted with a boss for an immersion heater.

Most hot water cylinders these days are copper, either uninsulated (though you should fit your own insulating jacket) or pre-insulated with a thick layer of polyurethane foam. The pre-insulated kind is much better at keeping the water hot and only costs a little more.

The standard size of copper hot water cylinder is 458mm in diameter and either 915mm or 1066mm high, holding around 140 litres. Other diameters include 300mm, 406mm and 508mm; other heights include 1220mm, 1525mm and 1828mm. The 1525mm × 300mm size is particularly useful for replacing a rectangular tank, since a standard-size won't fit if the tank is in a cupboard. It is generally the tallest made in a pre-insulated version.

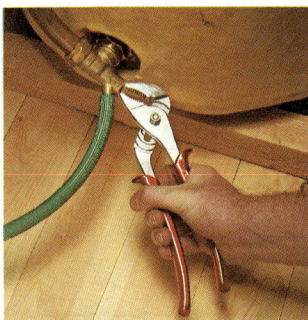

1 *Using an adjustable spanner, drain down the cylinder through a length of hosepipe*

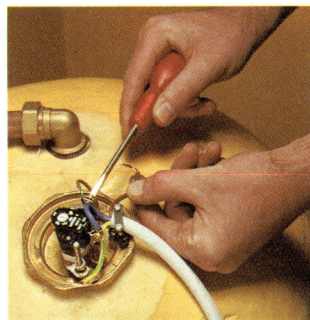

2 *Switch off at the mains before disconnecting the heater flex from the thermostat*

3 *Apply PTFE tape or plumbers' putty to seal the new immersion heater threads*

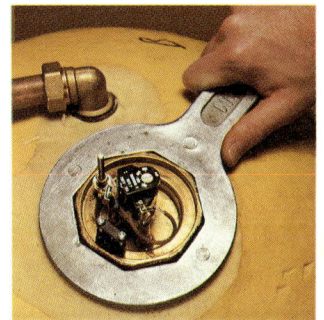

4 *Screw the new immersion heater to the tank boss tightly using an immersion heater spanner*

As well as choosing the size and type, you will also have to choose between a direct and an indirect cylinder, depending on what hot water system you have or want to have installed.

In addition, you will have a choice of where the bosses are for fitting immersion heaters—most cylinders on display in shops have a top fitting designed to take a vertical immersion heater.

Kits are available for converting direct cylinders to indirect ones and for fitting immersion heater bosses if your cylinder doesn't have one.

plus a drain cock at the cylinder, and that this pipe should feed *only* the hot water cylinder. You will need the appropriate number of 22mm elbows and 'T' fittings plus pipe clips to support the pipe, and expanded foam lagging to protect any cold water pipes in the loft.

Typical layouts are shown in the drawing above left. All hot water pipes must be taken off the vent pipe *above* the level of the cylinder and the pipes must fall slightly away from the vent pipe to avoid air locks.

The tools you need will be the same as for installing an immersion heater, plus a large

Spread some old sheets on the floor to catch any spills of dirty water. Old tanks can be very heavy—particularly the rectangular galvanized kind—so get help to carry them downstairs.

Fitting the new tank

Once the old tank has been removed, run any new pipes that are necessary for the cylinder. If you're running a new cold water feed, the cold water cistern must be drained so that the pipe can be connected into it.

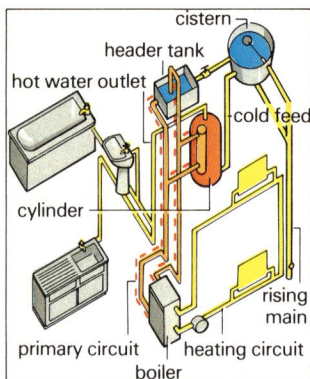

A separate electrical circuit should be run from the consumer unit to power the immersion heater

1 *If the drain cock is damaged or rusted solid, siphon out the water with a hose*

2 *After laying down drip sheets around the tank, disconnect all the pipes using a Stilson wrench*

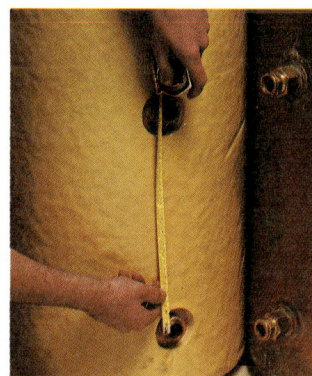

3 *Compare the position of the pipe connections on the old cylinder with the new ones*

What you'll need

The different types of hot water cylinder are described on page 101. If you're going for a replacement, it would be worth trying to find a cylinder that has its pipe connectors in the same place to avoid having to alter any of the existing pipework.

If you're putting in an indirect cylinder to replace a direct one, make sure that the connections will be accessible when the cylinder is in place.

If the cylinder is heated only by an immersion heater, it would still be worth fitting an indirect cylinder so that a central heating boiler connection can be made in the future. Make sure that the new cylinder will fit in the place taken up by the old one (allowing for insulating jacket or pre-insulation) and that there is adequate support.

If you need to run any pipework for either the hot water supply from the cylinder or the cold water supply to it, stick to 22mm tube—it will give much better flow rates than 15mm and will help to avoid air locks.

Remember that there should be a gate valve on the cold supply from the cistern

plumber's wrench if you're removing an old rectangular tank or any direct cylinder, since the joints will be very stiff to undo (hire a wrench if you haven't got one—900mm isn't too big for this job). You'll also need the normal plumbing tools if there's any pipework to renew—pipe cutters, spanners, a blowlamp (if you're using capillary fittings), a bending spring (or, for 22mm pipe, a hired pipe bending machine) plus a tape measure.

Removing the old tank

For this job, the entire cylinder will have to be drained down and, if necessary, the pipes back to the boiler (for a direct system). Take the hosepipe from the drain tap outside if possible and proceed as for replacing an immersion heater.

Removing an old tank can be strenuous work and you may have to take measures such as sawing through pipes with a hacksaw if you can't get the old, corroded nuts undone: if you weren't planning to change the pipes this will, of course, mean that you now have to.

After you have taken the old cylinder out, and before you have installed the new cylinder, stand the cylinders against each other and measure—from the centre of the holes—the distance between the primary circuit flow and return.

When you install the new cylinder, offer it up first and arrange it so that drain cocks, immersion heater bosses, gate valves, and all the other things to which you may need access are easy to get at. The old pipework needn't necessarily be moved to suit the new cylinder: simply use flexible copper pipe to join the pipework and cylinder. Be careful to keep bends in the tubes gradual, to prevent air locks.

If putting in an immersion heater for the

★ WATCH POINT ★

You'll need to do a couple of dummy runs first to ensure that the pipework goes together properly, before installing the cylinder for real. As you fit each pipe, label it clearly for identification later.

first time, run the electric wiring at this stage. If you're not having one, check that the blanking plate is screwed tightly home—some plumber's putty or PTFE tape will stop it leaking.

Put the new cylinder in position on its platform and connect the pipes—again using plumber's putty or PTFE tape on the indirect screwed connections to and from the boiler if they're being used. The cylinder can now be filled and the immersion heater connected to its switch, following the procedure described in Replacing an immersion heater (page 100), and the

boiler. A horizontal heater can be positioned below these, a vertical one at the top and as near the centre as possible so that it points down the middle.

If the cylinder has foam insulation, cut this away with a handyman's knife first.

The hole you need is fairly large and can be cut in one of two ways—either using a hole saw or tank cutter fitted to an electric drill, or by making a series of small holes in a circle and then removing the waste and filing the large hole to shape. If you're adopting this method, it might be better to remove the cylinder completely to avoid

fitting the heater and, if necessary, replacing the cylinder.

Insulation and scaling

Whatever method you use to heat your hot water, there's little point in allowing it to cool down before you can use it. An insulating jacket should be fitted to all plain copper cylinders, and it won't hurt to lag hot pipes within the airing cupboard as well, using slip-on lengths of foam pipe insulation tied or taped securely at bends to keep it in place.

4 *When fitting flexible copper pipe bend the pipe gently, a little at a time, to avoid fractures*

5 *Fix the switch to control the immersion heater in a place where you can turn it on and off easily*

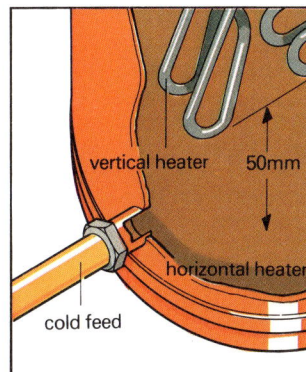

Ensure the immersion heater is at least 50mm above the base of the clinder to guarantee safe operation

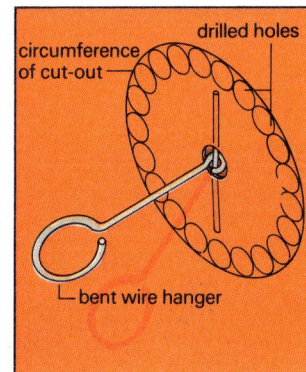

A clean hole can be cut in a cylinder using an ordinary drill but it is laborious work. It is easier to use a hole saw

system inspected for leaks before turning on the electricity or lighting the boiler.

Don't forget to fill the primary circuit if you have an indirect boiler, which may mean bleeding the radiators if it is connected to the central heating system—see page 89.

Cylinder conversion

If you want to put an immersion heater in a tank that hasn't got a hole, or you want to put a new one in a different position in the tank, you'll have to make a new hole for it (there's a standard size for this in the UK—2¼in. BSP). After the cylinder has been drained, the job consists of two parts: making the hole and fitting the boss.

Making the hole

If you have an indirect cylinder, you will have to be careful about where you position the hole for an immersion heater. Do not drill it in the space between the two tappings for the primary flow and return from the

getting a lot of copper filings inside it.

With the drill-and-file method, you'll have to drill a small hole in the centre of the cut-out and secure it with a piece of bent stiff wire to stop the 'waste disc' falling inside the cylinder. If you're using a hole saw, stop cutting when you're almost through the cylinder wall and prise out the disc as if you were opening a tin, bending it to break the remaining metal. Remove all the waste and clean around the outside of the hole.

Fitting the boss

The boss with all its necessary washers (and a template for making the hole) comes in a complete kit with full fitting instructions. There are different versions, depending on whether the boss is to be fitted on the side of the cylinder or on the top domed section—make sure you get the right one.

When fitting the boss, hold it securely with a length of wire to prevent it falling inside the cylinder as you tighten the whole assembly up.

When you've fitted the boss successfully, follow the procedure already described for

Hard water problems

If you live in a hard water area, two problems will arise to reduce the efficiency of your immersion heater and ultimately to shorten its life. Scale will build up comparatively quickly inside the cylinder and on the immersion heater itself, and if this build-up becomes too thick the heater will need to be on for longer and longer periods. You can reduce the incidence of scale by setting the heater thermostat to no more than 60°C.

Hard water can also corrode the heater element's cover, leading to early failure. If this is a problem in your area, fit a special corrosion-resistant type of heater which will have a far longer life. As a long-term aim, you could consider fitting a water softener.

Other alternatives for dealing with hard water include a **chemical scale inhibitor** (a bag of chemicals suspended in the water tank) which will stabilize the water, and a **magnetic conditioner** which generates a strong magnetic field and so counteracts the effect of impurities in the water. A **water filter** eliminates many of the problems associated with hard water.

INDEX